T0252557

Routledge Revivals

Progress in Rural Geography

This wide-ranging volume, first published in 1983, reflects the increasing scope of the field of rural geography in the second half of the twentieth century. Although traditional areas of study such as agriculture and the land-use patterns of the countryside remained important, scholars also began to consider rural transport, employment, housing and policy, as well as to develop new theories and methodologies for application to study. The chapters included here addressed the need for a review of the changes that had taken place within the field of rural geography, and as such provide an essential background to students with an interest in rural demography, planning and agriculture.

Progress in Rural Geography

Edited by

Michael Pacione

Routledge
Taylor & Francis Group

First published in 1983
by Croom Helm Ltd

This edition first published in 2013 by Routledge
2 Park Square, Milton Park, Abingdon, Oxon, OX14 4RN

Simultaneously published in the USA and Canada
by Routledge
711 Third Avenue, New York, NY 10017

Routledge is an imprint of the Taylor & Francis Group, an informa business

© 1983 Michael Pacione

All rights reserved. No part of this book may be reprinted or reproduced or utilised in any form or by any electronic, mechanical, or other means, now known or hereafter invented, including photocopying and recording, or in any information storage or retrieval system, without permission in writing from the publishers.

Publisher's Note
The publisher has gone to great lengths to ensure the quality of this reprint but points out that some imperfections in the original copies may be apparent.

Disclaimer
The publisher has made every effort to trace copyright holders and welcomes correspondence from those they have been unable to contact.

A Library of Congress record exists under LC control number: 82022756

ISBN 13: 978-0-415-70709-1 (hbk)
ISBN 13: 978-1-315-88668-8 (ebk)
ISBN 13: 978-0-415-70760-2 (pbk)

PROGRESS IN RURAL GEOGRAPHY

Edited by Michael Pacione

CROOM HELM
London & Canberra

BARNES & NOBLE BOOKS
Totowa, New Jersey

© 1983 Michael Pacione
Croom Helm Ltd, Provident House, Burrell Row,
Beckenham, Kent BR3 1AT

British Library Cataloguing in Publication Data

Pacione, Michael
   Progress in rural geography.
   1. Rural geography
   I. Title
   910'.091734   G115
   ISBN 0-7099-2021-0

First published in the USA 1983 by
Barnes & Noble Books
81 Adams Drive
Totowa, New Jersey, 07512

ISBN 0-389-20358-0

Typeset by Leaper & Gard Ltd, Bristol
Printed in Great Britain
by Billing and Sons Ltd Worcester

CONTENTS

TO MICHAEL JOHN

FIGURES

TABLES

Tables

PREFACE

Rural geography has a long tradition but until recently it generally referred to studies concerned with agriculture or comprised historical analyses and descriptions of the settlement or land-use patterns of the countryside. Although these areas of investigation retain their importance within the subject today, rural geography has expanded over the last decade to encompass other lines of enquiry such as the systematic study of rural transportion, employment and housing; assessments of development policies in rural areas; and attempts to develop theory and methodology in rural studies. The range and volume of work carried out under the umbrella of rural geography during the 1970s generated considerable enthusiasm for the subject and the momentum built up has ensured a central position for this branch of the discipline in the 1980s.

A result of this growing popularity and research activity is that scholars face a major task in maintaining contact with recent developments, keynote statements and relevant articles spread across a wide range of professional journals and less accessible reports. This volume is a direct response to the need for a text which reviews the progress and current state of the subject and which provides a reference point for future developments in rural studies. This collection of original essays is designed to encapsulate the major themes and recent developments in a number of areas of central importance in rural geography.

Michael Pacione
University of Strathclyde
Glasgow

INTRODUCTION

A full understanding of contemporary change in the built countryside cannot be gained without reference to the historical development of the settlement pattern. In Chapter 1 Michael Bunce analyses the evolution of the rural settlement pattern in Europe and North America. He first reviews several taxonomic approaches to the study of rural settlement and provides a critical examination of the nucleation-dispersal model, before specifically considering the influence of economic factors on the location and distribution of rural settlements. The importance of external forces, such as the rise of commercialism and urbanism in the post-medieval period, on the growth of rural market centres and, more recently, the adjustments forced on rural settlements by the scale and pace of urbanisation and industrial development are emphasised. He concludes that the plethora of theoretical and generalising ideas which have been proposed over the past century or more has identified the basic processes and factors which govern the development of settlement patterns. These include the influence of environment and culture on location and form; the fundamental relationship between settlement, resources and land use; and the competitive forces of the space economy. However, he cautions that this represents only a partial explanation, for it embodies a high level of generalisation and implies a degree of determinism which is not always appropriate for the interpretation of the complex adaptations of general factors to local conditions.

The growing pressure on rural land arising from the extension of urban influences, and the increasing competition among traditional land uses are matters of current concern. In Chapter 2 Tony Champion bases his discussion of rural land-use competition on the landscape model devised by Alice Coleman. He first considers the progress made in describing and measuring the composition of land use and patterns of land-use change for the whole system, before reviewing attempts to forecast land budget trends. Irrespective of whether the land budget can be balanced at a national level, conflicts between urban and rural interests can and do arise at a more localised scale. This is particularly the case in the 'urban fringe' and this zone is the subject of detailed examination. The focus of attention is then transferred to the 'marginal fringe' and the particular problems of this area at the outer edge of the

1

productive rural area. Finally the conflict between agriculturalists and conservationists in the intervening zone or 'farmscape' is examined. Several topics are identified for further research, including cost-benefit analyses of alternative levels of rural-urban land conversion; investigation of the compatibility of different land uses; and the monitoring of management experiments aimed to minimise land-use conflicts.

Farm structure, or the size and spatial arrangement of land holdings, is a fundamental factor in agricultural production. In Chapter 3 Ian Bowler analyses the key processes of land consolidation and farm enlargement in developed economies. He begins by examining the socio-cultural, economic and physical causes of land fragmentation and discusses the measurement problems attached to international study of the phenomenon. He illustrates that at a low level of economic development advantages may be realised by such an arrangement of holdings but once a commercially-oriented agriculture develops fragmentation increasingly imposes economic costs on farming. The process of land consolidation is then examined and the progress of government schemes in a number of countries is evaluated. In the second part of the chapter a similar interpretive structure is applied to analyse the phenomenon of farm enlargement. Finally several issues are pinpointed for further investigation. These include the need for more sophisticated techniques to measure the spatial patterns of land and farm fragmentation; an international survey of the progress made by consolidation schemes; examination of the causes of the spatial variation in the spontaneous process of farm enlargement; study of the agricultural costs and benefits of farm fragmentation; and assessment of the social consequences of structural change in agriculture.

In Chapter 4 Andrew Gilg discusses rural population trends and characteristics in general before presenting a detailed study of Great Britain, a country with one of the oldest census records. Data from the 1981 census confirm that regional variations in population growth or decline first noted in the 1960s continued into the 1970s. Thus the South West and East Anglia remain the fastest growing areas outside the rural South East, mainly due to a flight from cities and to restrictive planning policies preventing the spread of conurbations. More detailed analysis at the county, district and local levels provides further evidence of new trends in the distribution of the rural population. Explanatory models of rural population structure based on the concepts of population density, population potential and multi-variate classification are discussed before the focus of attention is directed to the related theme of rural employment. Despite the restricted and ambiguous nature of employment statistics it is clear that employment in British agriculture has fallen for over a hundred years and continues to fall annually. The suburbanisation of the countryside around towns has meant that the real employment problem is to be found in the remoter and upland areas. Following an examination of the nature of this dilemma and an assessment of possible solutions it is concluded that a co-ordinated approach by public bodies and the encouragement of locally-based enterprises are essential elements in attempts to alleviate the severe problems facing rural economies.

Until recently geographers afforded little attention to the study of rural housing, with most early research on the subject pursuing the link between house type and regional economic and cultural characteristics. Since the Second World War, however, rural housing in advanced capitalist societies has increasingly become divorced from the agricultural economy to which it was previously tied. Housing in the countryside now has to serve a variety of needs arising from a broadening employment structure and from the demands of recreation activities and retirement, in addition to its traditional function of providing shelter for the agricultural workforce. As a result the geography of rural housing has become as multi-faceted as its urban counterpart. In Chapter 5 Alan Rogers reassesses the role of housing in the rural landscape and reviews the contemporary situation in Western societies. He first examines the structure of rural housing and, using statistical evidence from France, Britain and the USA, demonstrates how the stock is generally older and in poorer condition than in urban areas. He emphasises the fact that contrasts in housing quality do not exist in a vacuum but are related to social and economic structures which constitute national and regional economies. Social and spatial variations in rural housing provision are discussed, and specific factors which underlie these patterns — household income, occupation and, notably in the USA, minority groups — are examined. The impact of the leisure society on rural housing is then considered before attention is given to the relationship between rural housing and planning policy.

Concern over levels of accessibility in rural areas has increased during recent decades. The gradual concentration of educational and health-care facilities and the closure of village shops, post offices and petrol stations have resulted in a decline in local accessibility and an increase in the need for transport. In Chapter 6 David Banister first defines the concepts of transport mobility and accessibility and then examines the dynamics of change in public and private transport usage. Changes in transport policy at a national level clearly have an important

impact on the provision of services in rural areas, and detailed consideration is given to the effects of relevant Acts of Parliament and in
particular to the recent Transport Acts. The alternatives available to
the planners and transport operators are considered, including increased
government subsidies, higher fares, route rationalisation, and the
transfer of operations to private companies. It is concluded that the
role of conventional public transport in rural areas may be limited. A
number of transport-related proposals, such as dial-a-bus, village cars,
post-buses, mobile services, as well as several non-transport options for
easing the problem of rural accessibility are considered. The fundamental principle which is emphasised, however, is the need to adopt a
broad perspective that includes all aspects of rural life to determine
whether the problem of accessibility is primarily a transport one.

The related themes of social change and deprivation form the basis
of Gareth Lewis's discussion of rural communities in Chapter 7. He
rejects the idea of the rural-urban continuum as a useful organising
concept and replaces it with a time-space framework which is not only
specifically rural but which considers the rural community as part of a
wider social and spatial system. Rural social change and recent population movements are interpreted within this context. Particular attention
is given to the 'population turnaround' experienced in some countries
of the developed world and the effect this process has had on the socio-
spatial structure of rural communities. The impact of public policy on
the changing rural community is also discussed. He points out that the
rural population has always suffered hardships of some kind but since
the Second World War diminishing employment opportunities, depopulation, and the declining levels of service provision have accentuated
the difficulties of country life. Attempts to measure the quality of rural
life are discussed and the important component of 'availability of
service provision' is examined in detail. It is suggested that for a more
effective policy to emerge greater understanding is required of the
spatial manifestation of service decline and of the motivations underlying the consumer movements of various rural groups. Three main
issues are recommended for further investigation: the degree to which
elements of deprivation are specifically rural in character; the extent
to which deprivation manifests social change; and, in order to help
understand the contemporary and future rural community, studies of
the individuals, groups and institutions who control the allocation of
resources.

Since the early 1960s there has been a major expansion of outdoor
recreation opportunities throughout the developed world reflecting

the pressures of demand stimulated by growing prosperity, and there are now many different types of recreation resource in most industrialised countries. In Chapter 8 Michael Tanner examines the emergence of problems related to the increasing pressure on rural recreation space. He illustrates how demand-based studies in the 1960s commonly led to attempts to improve the planning machinery and to increased allocation of public funds for recreation developments; a trend which marked a departure from the resource-based planning that had characterised the evolution of national park systems. While the lack of reliable forecasting techniques means the future for outdoor recreation cannot be predicted precisely it is likely to remain an important user of rural land. It is suggested that whereas the earlier developments in North America and Britain were based on the national park idea the emphasis in the future is likely to be increasingly on the provision of intensive-use areas designated to cater for the immediate recreational needs of a predominantly urban population, and on local recreation trails and footpaths. Another key issue centres on the need to minimise visitor impact on the resource base, and it is emphasised that the maintenance of a landscape suitable for recreation will depend primarily on the development of effective systems of multipurpose management.

In Chapter 9 Paul Cloke defines rural resources as those social and economic life-style opportunities where differential access by various income and age groups can lead to the problems collectively termed deprivation. He illustrates how rationalisation processes underlying deprivation have very often been directly and causally linked with the management of rural resources. Particular attention is given to the extent to which the planning system has been able to derive socially and politically acceptable policies for rural resource allocation *and* to implement these policies successfully. It is clear that equitable resource allocation policies, however evaluated, will not be sufficient to tackle the problems of rural disadvantage if there are difficulties in policy implementation, and the reasons for the gap which often exists between policy formulation and implementation are analysed. A fundamental question is whether sufficient resources exist, either within rural areas or in the proportion of national resources presently allocated for that purpose, to enable a positive response to rural problems. Possible approaches based on the use of existing resources and on the attraction of new resources are examined. It is concluded that while many of the alternatives discussed would require major alterations to existing philosophy such radical steps may be necessary because of the progressive trends of imbalanced resource distribution and social inequality in rural areas.

Rural planning may be defined as a form of public intervention to ensure a coherent and stable pattern of land use in rural areas in such a way as to produce the most life-enhancing pattern of activity for the community as a whole. Planning systems thus represent both the systematic allocation of land to particular uses and the resolution of competing and conflicting value systems in society. In Chapter 10 David Robins provides an overview of the major issues currently of concern to rural planners. Four 'geographical type areas' are identified — the urban fringe, agricultural lowlands, hills and remote uplands — each with distinctive planning problems. The fact that the statutory land-use planning system and a medley of resource planning functions are at work concurrently leads to consideration of the institutional framework for rural planning in Britain. The development of the current planning system is examined and the consequences of the major Parliamentary Acts evaluated. Attention is then given to a range of issues, including experimental work in relation to National Park developments and countryside conservation; settlement policy in rural areas; rural deprivation and the need to introduce a social base to rural planning; the need to develop a managerialist model of rural decision-making; rural employment; housing; and planning control over agriculture. In conclusion it is suggested that the most crucial question in British rural planning in the 1980s is the extent to which the embryonic approach to the integrated management of the countryside can develop into a coherent planning system.

1 THE EVOLUTION OF THE SETTLEMENT PATTERN

M. Bunce

The recent revival of interest in the rural areas of the industrialised world is, understandably, dominated by the question of contemporary change. Yet the preoccupation with recent rural transformation frequently ignores the settlement framework within which it occurs. Rural settlement exhibits strong continuity with the past, creating landscapes which often outlive the recurrent social and economic changes that occur within them. This is not to say that the framework is static, rather that it has historical depth evolving, to use Hoskins's (1955) term, as a palimpsest: layers of evolution in which features of various times in the past contribute to the patterns of today.

The pattern of rural settlement in England is strongly influenced by significant periods and events. For example, the Anglo-Saxon settlement was responsible for clearing large areas of woodland and establishing the framework for village settlement and open-field cultivation. The Norman Conquest heralded the feudalism of the Middle Ages and entrenched village society as well as the concept of large rural estates which has continued to be so much a part of the English rural landscape. The Black Death led to widespread village abandonment. The enclosure movement added dispersed elements and created the diversified settlement pattern which exists to this day. More recent events, such as the rapid rise of industry and urbanism and the growth of land use planning, have all had a major impact on the pattern of rural settlement.

Taxonomic Approaches

The history of rural settlement, however, has unfolded in a variety of regional and local circumstances. The diversity of European and North American rural landscapes is the result of the unique combination of historical variables in each region and even in each place. It is the recognition of this that led to the tradition of local and regional description in rural settlement studies. Yet the need to establish some interpretive order out of this empirical complexity was quickly recognised. This began with the descriptive work of the German and French schools of settlement geography. The profusion of regional descriptions which dominated nineteenth-century settlement study soon produced a call for taxonomic order. Classification, it was presumed, would establish a framework which would permit general interpretations of rural settlement patterns. Strongly influenced by the German School and the later work of French scholars numerous typological schemes have been proposed. These have employed a variable mix of morphological, locational, functional and genetic criteria in a largely futile attempt to produce taxonomic models which could have general application.

However, any attempt to make sense of the complexity of rural settlement patterns must begin with the summary of its main components, and in this sense, classifications have been helpful. Most typologies are based upon the division of rural settlement into nucleated and dispersed forms, with further subdivision into the three basic elements of village, hamlet and dispersed farmstead. Certainly, over the broad sweep of history, these have been the basis of rural settlement structure. However, this framework is not readily applied to a spatial dimension. Various attempts have been made to divide Europe into uniform regions of villages, hamlets and dispersed settlement. Thorpe (1964) for example, has proposed broad zones of rural settlement along these lines. Yet this, and other regional classifications, have little empirical foundation. Firstly, the spatial distinction between nucleation and dispersal is rarely clear-cut. Secondly, few regions can be said to be composed of a single settlement type. The process of settlement evolution has produced a mix of elements. Nucleated and dispersed forms, villages and farmsteads co-exist in most rural areas of Europe and North America, and, furthermore, are functionally interdependent.

Despite the mainly descriptive format of most classifications, the intent of the taxonomic tradition has been, for the most part, to produce models of settlement patterns and to suggest the main factors influencing their evolution. Meitzen (1895), for example, based his typology upon the four criteria of morphology, field systems, ethnic origin and relationship to the physical environment. Demangeon (1927) classified villages according to field systems and morphology, and proposed four categories of dispersed settlements based upon the chronology of dispersal. The most recent attempt at classification, that of an IGU Commission, has so far generated 66 criteria for explaining rural settlements in terms of topological and chronological position, degree of permanency, economic and social structure, size and morphology (Uhlig, 1972).

Nucleation and Dispersal

One model in particular has dominated this work: the nucleation-dispersal dichotomy. The influence of Meitzen in this respect is legendary, for it was he who initiated the idea that rural settlement patterns should be interpreted in terms of the relative distribution of nucleated and dispersed farms. He suggested a direct relationship between the type of agricultural system and ethnic structure, which in turn determined the existence of nucleation and dispersal. His deterministic association of village settlement with Germanic culture, and of independent farmsteads with Celtic origins was easily discredited as regional exceptions were produced to disprove the rule.

However, the emphasis upon nucleation and dispersal has remained. Meitzen set off a debate which stimulated an interest in the refinement of general interpretations of the two types of settlement pattern. The rejection of his ethnic hypothesis led directly to interpretations which stressed the deterministic influence of particular factors such as field systems, physical environment, defence and social structure. However, attempts to use this as a basis of a general theory of European settlement (such as Gradmann's (1929) steppenheide thesis, and Aurousseau's (1920) association of village settlement with scarce water supply) were simplistically deterministic and had little empirical foundation.

To some extent, the exigencies of defence help to explain nucleating tendencies. The village form itself owes much to the need for group security during the initial stages of land settlement, and to the subsequent need to protect village territory. In regions of long-term insecurity, too, other factors have often been secondary to defensive requirements. Large, highly-concentrated villages in the Mediterranean region owe their origin to recurrent periods of instability. In Corsica, for example, the rural settlement pattern is dominated by strongly nucleated villages which are a response to what Thompson (1978) has described as 'a history of almost incessant conflict'. Again, however, we must be cautious of a deterministic interpretation. Agnew (1944-6) has pointed out that despite the great unrest of Languedoc between the end of the Roman period and the seventeenth century, the concentrated village was slow to evolve.

The role of physical factors in the development of a nucleated-dispersed dichotomy is somewhat more obscure than that of defence. Thorpe (1964) among others, has attempted to draw a clear distinction between upland and lowland Britain, arguing that upland environments encouraged pastoralism and therefore dispersed patterns, while the lowlands provided the extensive and fertile land base necessary for village settlement. Yet, many parts of upland Europe have long been dominated by villages, largely because the terrain leads to isolated pockets of close-knit communities which tend to favour village settlement. On the other hand, it can also be shown that the scarce and fragmented distribution of agricultural land in mountainous areas mitigates against large-scale village settlement and permits only scattered farmsteads.

The problem with considering the role of external factors such as defence and environment is that they do not operate in isolation, but in the context of social and economic circumstances. In an agricultural society this suggests that rural settlement patterns have a general relationship with agrarian systems, and in particular with the organisational framework of rural society. Meitzen himself proposed a fundamental relationship between field systems and settlement types, a hypothesis which has since received a good deal of empirical support. In terms of nucleation and dispersal, there is general agreement that village settlement has evolved in association with collective systems of open-field agriculture, while dispersal is linked to independent cultivation of enclosed fields.

Thus we can trace the evolution of the village in Europe to the establishment of first communal and then feudal systems of land tenure and use. Whether social organisation, economic necessity or environmental constraints caused this relationship between settlement and land use would have varied from one region and one period to another. For example, Saxon and Scandinavian settlers brought with them to Britain social organisations which favoured nucleated settlement. Yet the process of extensive land clearance and the ploughing of heavy soils reinforced the need for communality and cooperation and entrenched the village as a settlement form. With the rise of feudalism in Europe during the Middle Ages this structure was ensured to the advantage of land-owning groups, and the rural population became firmly fixed in village territories. The open-field system which accompanied most feudal arrangements demanded strict observance of cultivation laws and restricted the dispersal of peasant dwellings. Indeed, it has been suggested that in a system with scattered strips and compulsory crop rotation, the optimum settlement pattern would be a nucleated one (Smith, 1967).

There was considerable variation in the nature of feudal villages, particularly in the relative freedom of sections of the peasantry which seems to have been a function of the degree of manorial authority. Nevertheless, as Smith has suggested, the bulk of today's villages in Western Europe were founded by the mid-fourteenth century, under a political system which inevitably led to the nucleation of population into highly-organised agricultural communities.

By contrast, the dispersal of settlement has generally been associated with individual cultivation and tenure. There is considerable evidence to support this both in the original settlement of certain regions and in the dispersal of settlement from village cores. The dispersed patterns of parts of Ireland, Wales and Brittany, for example, owe much to the limited amount of community organisation of early Celtic society in which freemen lived outside small clan hamlets on independently cultivated land in a loosely-organised infield-outfield system (Johnson, 1961). Dispersed settlement was characteristic also of the expansion of settlement on to poorer land in parts of Europe during the Middle Ages. In areas such as the English Fens, the Pays de Caux and northern Sweden, scattered farmsteads originated in late-medieval piecemeal clearing and reclamation, the gradual extension of pastureland beyond village boundaries and the creation of enclosures by individual peasants.

The analysis of early dispersal of rural settlement in Europe favours both culture and economics as the motivation. Thus Celtic social organisation permitted the independent location of farms while the pastoral economy was more efficiently pursued within an infield-outfield system. And the colonisation of new land was the result of land shortages particularly at a time when manorial authority was beginning to loosen its control of the peasantry.

More extensive dispersal in Europe, however, is linked to the gradual weakening of feudalism and to the spread of enclosures in which the economic and social advantages of the individual farmstead set in its own fields were recognised. Herein lies the impossibility of distinguishing between regions of nucleation and dispersal, for over much of Europe the characteristic pattern is a mixed one: the result of various periods of secondary dispersal from village cores. The close association between independent cultivation and dispersed settlement reaches its height in the New World where early settlers eschewed the village and established the pattern of independent, dispersed farmsteads which so dominates the contemporary rural landscape.

In Pennsylvania, for example, William Penn's original plans for settlement were for an orderly scheme in which townships would be dominated by agricultural villages. Yet the general pattern that developed after 1700 represents a direct rejection of Penn's philosophy, by settlers who held a quite different view of the New World. Their ideology was an independent, liberal one which represented a desire to break away from the already declining authority of the European village. Some groups, such as the Quakers, had religious reasons for avoiding Penn's villages, but most settlers saw both economic and social advantages in living at the centre of their newly-acquired land (Lemon, 1972). In Quebec too, the feudal seigneurial land division system in the early seventeenth century was not transferred intact from France. As Harris (1966) has shown, the independence of the *habitants* (settlers), the general poverty of the *seigneurs* (manorial lords), and the pioneer environment of the St Lawrence lowlands limited the social and economic influence of the *seigneurs*. Villages did not develop to any great extent, despite the wishes of the authorities, for the *habitants* immediately saw the benefits of independent settlement particularly when illegal participation in the fur trade offered useful remuneration.

These trends reflected the growing movement in Europe towards dispersal and greater independence of land use and tenure. Yet, like the enclosure movement itself, it represented a fundamental change in the relationship between settlement and land, in which the objectives of agriculture shifted from subsistence to commercialism. In some areas and periods in Europe, such as during the secondary dispersal of farmsteads in southern France and the Tudor enclosures in Britain, this involved local and private initiative. However, later enclosures in Britain, settlement schemes in Holland and, more extensively, the spread of settlement across North America, represented official policy on the appropriate pattern of rural settlement. In North America, the westward expansion of settlement was preceded by a survey which laid out individual lots within a grid pattern, a system which ensures a dispersed pattern of farmstead settlement.

Location and Distribution

Clearly the functional relationship between social structure, agrarian systems and settlement forms has strongly influenced the evolution of rural settlement. Yet a serious weakness of this as an explanatory model is that it has been framed largely in terms of nucleation and dispersal. The problem with this dichotomy is that it ignores the great complexity of settlement patterns. Even if we accept the notion of two great divergent forces in the history of rural settlement, actual patterns do not break down so simply. One important reason for this is that patterns are more a function of the location and relative spacing of elements than of their simple categorisation.

This has been recognised by recurrent attempts to explain rural settlements in terms of the factors which have influenced the specific location of elements. Contemporary patterns to some extent reflect site and locational decisions taken at various times in the past. Again the roles of defence, of the physical environment and of social and economic factors have been emphasised. However, much of the work along these lines has been highly descriptive, interpreting each factor in a single cause and effect relationship. Clearly the selection of settlement sites would have involved a combination of variables the relative importance of which varied from place to place.

Chisholm (1968) has approached this problem by proposing a systematic model in which the siting of settlements is determined by the availability and relative importance of agricultural land, water supply, fuel and building materials. The site chosen for settlement would be that which achieves the minimum total cost of using these resources, rather than one which is attracted to a single resource. Yet Chisholm also recognises that settlement patterns are not simply determined by initial site-selection but by subsequent processes which affect the spacing of individual settlements. Noting the apparent regularity of the spacing of villages in eastern England, Chisholm suggests that the extent of village territory, in part, would have been determined by the maximum distance that villagers were prepared to travel to their parcels of land. Beyond that, new villages would have been formed whose spacing was a function of the optimal size of each village's territory.

The same principle can be applied to patterns of dispersed farmsteads in which the spatial distribution would be a function of the maximum economic distance between the farmstead and the furthest fields. In reality however, farmers have rarely been free to optimise either farm size or location in this way. In the dispersal of farmsteads in Europe both were generally controlled by the size of enclosed fields and the size of farms established by landlords. In North America, the various survey systems very much determined the size of farm holdings. Only within these was the farmer free to choose farmstead location.

Chisholm's thesis is one of a number of attempts to identify systematic relationships and processes in the evolution of settlement patterns. Within the framework of diffusion theory, Bylund (1960) and Hudson (1969) have proposed models in which the spread of settlement across a region proceeds by a series of stages, the location of each settlement being determined by the location of those of earlier stages of development. This clearly has some merit, for most settlement patterns have evolved through a process of diffusion of both population and ideas. Hudson, however, recognises that once the settlement of a region is established the subsequent evolution of the pattern proceeds through the process of internal adjustment, an important element of which is competition between settlements. This is particularly relevant once rural society develops beyond the stage of agrarian subsistence, and villages and hamlets take on the role of service centres for a commercial farm population. This is the point at which Christaller's and Losch's work on settlement hierarchies becomes relevant. The basic hypothesis of central place theory is that competition between places for consumers will result in an hierarchy of service centres which will in turn determine, through the size of their market areas, the spacing of settlements of different sizes (Berry, 1967). Christaller observed such a hierarchy in southern Germany, although the regularity of pattern proposed by his theory is not characteristic of most of rural Europe. However, it does explain the principle behind the process of settlement rationalisation which has gone on since as far back as the Middle Ages, in which continued competition in the rural space economy has resulted in the growth of larger centres at the expense of smaller ones.

Empirical tests of central place theory, however, have shown it to be particularly relevant to the evolution of the pattern of nucleated settlement in much of North America. The relatively recent settlement of the American Middle West and the Canadian Prairies involved a fairly evenly dispersed farm population and the establishment of many small trade centres. The process of competition between these places has produced a hierarchical and largely regular pattern of small towns, hamlets and villages.

The plethora of theoretical and generalising ideas which have been proposed over the past century or more has identified the basic processes and factors which govern the development of settlement patterns: the influence of environment and culture on location and form, the fundamental relationship between settlement, resources and land use, and the competitive forces of the space economy. Yet we must remember that this represents only a partial explanation, for it embodies a high level of generalisation and implies a degree of determinism which is not always appropriate for the interpretation of the complex adaptations of general factors to local conditions.

Commercialism and Urbanisation

Of more significance, however, is the tendency for most work on settlement patterns to emphasise the processes involved in the initial establishment of settlement and to see subsequent development in terms of the internal evolution of rural society. Yet rural settlement, as we have seen, has been very much affected by external factors. To a large extent these have represented the growing dominance of the city over the country. In Europe it began with the rise of feudalism, which was imposed upon existing agricultural communities by a land-owning class, primarily interested in the political control of population and the raising of revenue for non-rural institutions. Village expansion, the dispersal of farmsteads, the enclosure of open fields, the reclamation of marginal land, all of which form the backbone of many of today's rural landscapes, were associated with an expansion of the urban population and the rise of commercialism in the post-medieval period. Two of the most lasting effects of this were the establishment of villages and small towns as market and service centres, and the spread of the great country estates. The former evolved as a key element in the rural settlement pattern, providing a permanent point of exchange between the agricultural and urban economy and a focus for the investment of capital. During the Tudor period, many English villages ceased to be merely peasant communities as the combination of rural industry and commercial agriculture played an increasingly important role in the village economy (Thirsk, 1961). This pattern was repeated elsewhere in Europe whenever capital was available, with many larger villages growing into the market towns which served as the economic and administrative centres of rural regions.

The country estate extended the dominance of rural by urban society, with the ownership of large sections of rural land by both the nobility and the bourgeoisie. Apart from the obvious impact of the great houses and their grounds, the most significant influence on the settlement pattern was the system of tenancy which permitted the landowner to determine the nature and extent of development in both village and farm. This led to considerable village planning and reconstruction, and farmstead re-development. It was accompanied during the seventeenth and eighteenth centuries by the spread of scientific and technological innovation to agriculture. This completed the triangle of land-ownership, capital investment and commercial production.

The commercialisation of the rural economy produced the mixed pattern of dispersed farmsteads, market towns and service villages which is the landscape of many parts of Europe today. It has not, of course, affected all areas. In parts of France and Germany, for example, the farmstead remained within the village and fields have yet to be enclosed despite the predominance of commercial agriculture. And in other areas, such as southern Italy, Spain, western France and the Balkans, rural society continued much longer within a largely feudal system and a locally subsistent economy, with little change in settlement patterns between the Middle Ages and the present day.

The changes which accompanied the growth of urban and commercial influence in rural Europe were, of course, grafted on to the existing pattern of settlement. By contrast, the settlement of most of North America reflected, from the outset, a commercially exploitative attitude to rural land. The settlement patterns of the rural South of the United States have been dominated by the history of a plantation economy, geared to the development of urban-based industry. Rural settlement in Newfoundland, New Brunswick and Nova Scotia owes much of its present character to a colonial need for timber and fish in the seventeenth and eighteenth centuries.

These were economic developments which coincided with the shift in urban-rural relationships which had been occurring in Europe. And this was extended in the spread of agricultural settlement across the continent. The settlement patterns of the United States Middle West, of Ontario and the Canadian Prairies reflect an official desire for order and equity in the occupation of new land. Yet underlying this was a concern for economic independence on the part of the settlers and for a commercially productive agriculture to serve the markets of the cities of the eastern seaboard and of Europe. The survey and land-allocation system, for the most part, was based upon the principles of the optimal land parcel size and settlement pattern for a successful agricultural economy. The pattern that evolved, therefore, was one of a dispersed farm population with nucleated settlements growing up as places which served the economic and social needs of that population.

The rural trade centre became a particularly important component of this landscape. In the spread of settlement along the Ohio Valley between 1800 and 1860, a network of 148 villages and small towns developed to service the frontier economy: entrepôt towns along the Ohio river, merchant and milling centres at the interior junctions of small streams and local roads, which were service centres for what must have been small hinterlands (Muller, 1976). Evidence from central place studies shows that, with the spread of settlement westwards, these types of places evolved into a hierarchical framework based upon differing levels of service provision.

Recent Trends

By the early nineteenth century in Europe, and by a century later in North America, the pattern of rural settlement had largely reached its peak of development. Yet it was becoming increasingly integrated into the broader economic system, and in the period since then its evolution can be explained only in terms of adjustments to the quickening pace of urbanisation and industrial development.

Over the longer term this has led to the rationalisation and dilution of the pre-industrial settlement pattern. Increased commercialisation and capital intensification of agriculture, accompanied by the exodus of farm labour and rural industry to the cities, has had the effect of weakening the economic and social base of rural areas. As early as the 1870s there is evidence of an increase in farm size with a decrease in total farm numbers in Britain (Ernle, 1912). Certainly since the 1930s this has been a major factor in the evolution of rural settlement on both sides of the Atlantic. In part it has involved a retreat from the more remote and environmentally marginal areas such as the Scottish Highlands, the Massif Central and the Appalachians. Yet, more generally, the decline in farm numbers has led to a thinning-out of the settlement pattern. The absorption of smaller farms by larger ones has reduced both population and building densities in open countryside. This is particularly apparent in areas of massive farm enlargement such as the Canadian Prairies and the American Great Plains, where the decline in farm population has precipitated the disappearance of other dispersed elements.

The decline in farm settlement has been accompanied by the reduction in the importance of market towns and villages to the local economy. Again, in marginal areas persistent depopulation has often led to the abandonment of whole settlements. But at a more general level, the problem in many farming areas is that of a reduction in the size and importance of service centres. Saville (1957) has shown how persistent depopulation in rural England since 1850 has contributed to a decline in the secondary rural population and threatened the survival of villages as economic and social centres. This experience has been repeated throughout rural Europe and North America, for many areas have inherited a network of hamlets and villages from a period when the rural economy was more locally oriented. Of course, the inability of many settlements to sustain their role as economic and cultural centres is due to more than just a decline in farm population. It is also a manifestation of competition within the broader settlement system in which smaller centres cannot compete successfully with larger ones. With increased mobility and the growth of mass consumerism the economic advantages of concentrating services into larger centres have become unavoidable.

In this way rural areas have been absorbed increasingly into the urban sphere. And if we consider very recent trends, it is this process which now has the greatest influence upon the evolution of rural settlement. On to patterns which, as we have seen, are composed of many historical layers, are now being grafted the structural and physical changes resulting from industrialised agriculture, exurban residential development and recreational activity (Bunce, 1982).

Modern agriculture requires the adaptation of farmsteads and farm layouts and imposes new structures upon rural landscapes. The spread of urban population into the countryside and the rapid growth of the rural non-farm population turns villages into housing estates, farmhouses and cottages into rural retreats, and converts farmscapes to the sprawl of small country estates. Many rural communities, as a result, are no longer part of a diversified local economy but exist as a convenience for an urban and mobile population. The impact of this is particularly apparent in the growing perception of the rural environment as a playground for urban society. Recreational activity transforms agricultural and fishing villages into second-home communities, farmscapes into public parkland, old country estates into country clubs. It also adds a variety of new commercial elements from hotels and restaurants to golf-courses and antique shops.

The very pace of change in rural areas has resulted in the growth of public policies aimed at managing the process of change. This will be one of the most important factors affecting the future of rural settlement. Policies for settlement rationalisation, resource development and land tenure reform are among the more influential forms of government intervention aimed at maintaining the viability of rural areas. Yet, at the same time, the traditional character of rural settlement is being protected from change by a variety of conservational policies. These are more extensive and influential in Europe than in North America. In Britain in particular, the tradition of countryside conservation is strong, and rural change can occur only within the framework of planning legislation which, at least in principle, emphasises the protection of open countryside, the preservation of village character and the maintenance of traditional designs and structures (Gilg, 1979). Yet the extension of urban planning principles to rural districts has led also to the standardisation of building design and settlement morphology.

Conclusion

When considered over the broad span of history, the evolution of rural settlement patterns poses considerable interpretive difficulties. General processes and factors can be recognised. Particular trends and events have had widespread impact. Yet there remains great geographical diversity and the differences between rural regions and places are, in many ways, more significant than the similarities. The conformity of modern society, and certainly the standardising effects of modern planning, may weaken this diversity. However, even these influences must operate within local and regional contexts in which settlement patterns are the product of the unique combination of historical, cultural and economic factors.

References

Agnew, S. (1944–6), 'Rural Settlement in the Coastal Plain of Bas Languedoc', *Geography*, vol. 29–31, 66

Aurousseau, M. (1920), 'The Arrangement of the Rural Population', *Geographical Review*, vol. 10, 223–40

Berry, B.J.L. (1967), *The Geography of Market Centres and Retail Distribution* (Englewood Cliffs, N.J.: Prentice-Hall)

Bunce, M. (1982), *Rural Settlement in an Urban World* (London, Croom Helm)

Bylund, E. (1960), 'Theoretical Considerations Regarding the Distribution of Settlement in Inner North Sweden', *Geografiska Annaler*, vol. XLII, 225–31

Chisholm, M. (1968), *Rural Settlement and Land Use* (London, Hutchinson)

Demangeon, A. (1927), 'La Geographie de l'Habitat Rural', *Annales de Geographie*, vol. 36, 13

Ernle, L. (1913), *English Farming, Past and Present* (London)

Gilg, A. (1979), *Countryside Planning: the First Three Decades 1945-76* (London, Methuen)

Gradmann, R. (1929), *Die Arbeitsweise der Siedlungsgeographie* (Berlin)

Harris, R.C. (1966), *The Seigneurial System in Early Canada* (Madison, Wisconsin University Press)

Hoskins, W.G. (1955), *The Making of the English Landscape* (London, Hodder and Stoughton)

Hudson, J.H. (1969), 'A Location Theory for Rural Settlement', *Annals, Association of American Geographers*, vol. 59, 365-82

Johnson, J.H. (1961), 'The Development of the Rural Settlement Pattern of Ireland', *Geografiska Annaler*, Series B, vol. XLIII, 165-173

Lemon, J.T. (1972), *The Best Poor Man's Country* (Baltimore, Johns Hopkins Press)

Meitzen, A. (1895), *Siedlung und Agrarwesen der Westgermanen und Ostergermanen*, vol. 1 (Berlin)

Muller, E.K. (1976), 'Selective Urban Growth in the Middle Ohio Valley, 1800-1860', *Geographical Review*, vol. 66, 178-99

Saville, J. (1957), *Rural Depopulation in England and Wales, 1851-1951* (London, Routledge and Kegan Paul)

Smith, C.T. (1967), *An Historical Geography of Western Europe Before 1800* (London, Longmans)

Thirsk, J. (1961), 'Industries in the Countryside' in F.J. Fisher (ed.), *Essays in the Economic and Social History of Tudor and Stuart England in Honour of R.H. Tawney* (Cambridge, Cambridge University Press)

Thompson, I. (1978), 'Settlement and Conflict in Corsica', *Transactions, Institute of British Geographers*, vol. 3, 259-73

Thorpe, H. (1964), 'Rural Settlement' in J.W. Watson and J.B. Sissons, *The British Isles: A Systematic Geography* (London, Nelson)

Uhlig, H. (1972), *Rural Settlements*, International Working Group for the Geographical Terminology of the Agricultural Landscape, vol. II (Giessen)

2 LAND USE AND COMPETITION

A.G. Champion

Recent years have witnessed an upsurge of interest and a change in emphasis in rural land-use studies in response both to changes in the rural environment and to changes in the way in which that environment is perceived by academics and the wider population (Cloke, 1980; White, 1981). Among the most dramatic of these changes has been the growth of outdoor recreation and the increase in public concern over the appearance and ecological structure of the rural environment. These can be attributed to many factors including the marked increase in affluence and mobility in most developed countries over the past two or three decades, the shift in recreational activities from town and coast into the countryside, and the general rise in concern over environmental quality which appeared to be heralding the evolution towards a post-industrial society in the 1960s. At the same time, attitudes towards primary activities have been shifting as new production methods, together with a general intensification of production, have brought major landscape changes and led to greater negative externalities. In particular, farming, once considered the guardian of the countryside and an activity which needed protection from urban pressures, is now regarded by some to be the main scourge of the rural environment and an activity on which various restrictions should be imposed. In studying rural land-use trends, therefore, it is now necessary to look beyond the competition between land-using activities and the particularly well-documented struggle between agriculture and urban development and recognise the existence of clashes between a range of interests within the rural environment itself.

These developments have stimulated an extensive literature about the patterns and processes of land-use change in the countryside. Much of this work is diffuse in its origins, coming from economists and ecologists as much as from geographers and planners, so it is fortunate that several recent publications have synthesised the developments in particular areas and can provide a good introduction. Rhind and Hudson (1980), for instance, outline methods for the collection and manipulation of land-use data, while Best (1981) describes the patterns of land use and documents their changes over time. The Countryside Review Committee (1978) outlines the types of problems faced by farmers, and Davidson and Wibberley (1977) and Gilg (1978a) both stress the increasing scope for conflict between the powerful sectoral interests in rural land and recommend that greater integration and co-ordination be provided by an overriding body. The background to the growing conflict between farming and conservation is examined by Davidson and Lloyd (1977), while the more recent statements by Shoard (1980) and Green (1981) offer two rather different planning responses. Blacksell and Gilg (1981) illustrate a discussion of these themes with case-study material drawn from their research in Devon, and the new series of *Countryside Planning Yearbooks* brings together from a wide range of sources a wealth of factual and review information relating directly or indirectly to the use of the land (Gilg, 1980, 1981).

Given the obvious breadth of the subject and the fact that other chapters in this book describe in more detail some of the main ingredients of land use such as agriculture, recreation and settlement, this review opts for a selective approach concentrating on the study of contexts in which land-using interests overlap or compete for space. The five topics highlighted in this chapter can be introduced most simply by relating them to the landscape model devised by Coleman (1969) and shown in its most basic form in Figure 2.1. The first section investigates the progress made in describing and measuring the composition of land-use and patterns of land-use change for the whole system, while the second reviews the attempts made to forecast land-budget trends and to assess their implications for land-use planning decisions, particularly those concerning the extent to which urban development (broadly represented here as the townscape) should be allowed to expand at the expense of the rural area. The next two sections deal with the traditional areas of land-use conflict and controversy, portrayed in Figure 2.1 as marginal zones between the three 'pure' landscape types and known more generally as the urban fringe and the physical margin, while finally the review turns to the emerging environmental concern over developments in the intervening zone (or farmscape) where agriculture has in the past tended to reign supreme on fertile land situated largely beyond the threat of large-scale urban penetration.

Figure 2.1: The Scape-fringe Model

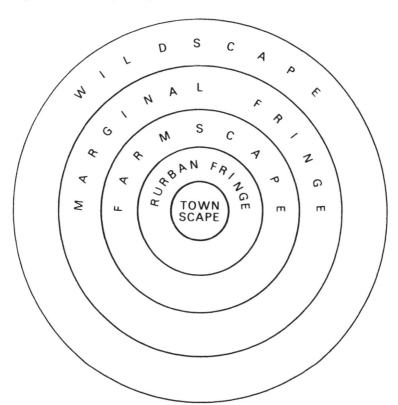

The Study of Land-use Patterns

One of the main areas of work on rural land use remains the identification and measurement of the land-use patterns themselves and of their changes over time (Cloke, 1980). As a result of this work the overall situation is reasonably well known for many developed countries and has been described by Best (1981), Best and Coppock (1962), Champion (1974), Clawson (1972), Coppock (1968, 1977a), Hart (1976), and Secretary of State for Scotland (1973). The main features appear to include the overall loss of rural land to buildings, roads and related urban uses, the sometimes more temporary transfer of land to mineral workings and other semi-urban uses, the afforestation of generally more marginal farmland and waste-land, and an increasing interest in rural areas for their recreational, amenity and ecological value.

The precise scale of these trends and their implications are, however, much more poorly understood. Though sampling procedures can substantially reduce measurement time (Fordham, 1974; Anderson, 1977), primary data collection has traditionally required vast amounts of manpower, even for a relatively small country like the UK, as accounts of its two nationwide surveys testify (Stamp, 1931; Coleman, 1961). Advances in remote sensing techniques and computerised data storage are beginning to overcome this problem (Rhind and Hudson, 1980), but the heaviest reliance is as yet still placed on secondary data sources. Since this data collection is usually undertaken by various agencies with specific interests in rural areas rather than by general bodies, it is not surprising that an accurate overall picture is difficult to obtain. Generally the most reliable, detailed and frequently updated statistics relate to the agricultural area, with woodland coming a poor second and with urban land being particularly badly treated (Best and Rogers, 1973, Clawson and Stewart, 1965, Coppock and Gebbett, 1978). The challenge presented to rural studies by deficiencies in data quality is compounded by the large areas that these sources jointly fail to cover, by smaller difficulties involving overlap between sources, and by the very great problems encountered in tracing land in multiple use (Mather, 1979).

The absence of land-use information which is comprehensive in coverage and regularly updated has allowed the development of misconceptions and 'myths' about this subject (Best, 1977). This is particularly the case for the rate at which rural, and particularly agricultural, land is being taken for urban development. Understandably, the scope for error is greatest in countries with extensive land areas, low pressures of population on land, and thus a relatively low concern over the land-budget implications of land conversion. According to Hart (1976), estimates of the annual average urban land-take in the USA in the 1960s range from 350,000 acres to 5 million acres. Even in the UK, despite its much higher overall man/land ratio and long-standing concern over land resources there remain considerable differences of opinion over the recent rate of agricultural-urban land conversion (Best, 1978; Coleman, 1978), though a much smaller margin separates Best's and Coleman's estimates of the urban land-take between the 1930s and the 1960s (Best, 1981).

An examination of the British case indicates the principal ways in which such discrepancies can arise. A key problem is that, because of the greater availability of agriculture and forestry data, the urban area has often been calculated as a residual, thereby ignoring the existence of other uses and waste-land and allowing the urban estimate to be affected by any errors in the other data. For example, an official estimate of the urban area for England and Wales in 1937 was over 50 per cent greater than the more accurate figure produced by the first land utilisation survey for the early 1930s (Best, 1959). A second major problem is presented by inconsistencies in land-use definitions, particularly in relation to urban land where the treatment of public open space and other semi-urban uses often varies between surveys. Thus, Coleman (1978) experienced difficulties in comparing the results of the second land-use survey undertaken in the 1960s with those produced for the 1930s. As pointed out by Rhind and Hudson (1980), it can also explain some of the discrepancies between the urban area estimates of Best (1976), Coleman (1978) and Fordham (1974).

The residual and definitional problems have sometimes combined to increase uncertainties over the rate of urban expansion. One major source of data on urban expansion, the Ministry of Agriculture's annual statistics on the amount of agricultural land taken for urban use, became less useful in the late 1960s when the Ministry adopted a policy of regularly redefining the agricultural area so as to exclude so-called 'statistically insignificant holdings' (Champion, 1975). Any such land that has been developed subsequently will have escaped the notice of this source, while conversely it appears that much farmland which may have been recorded as being sold for urban use may not actually be developed (Davidson and Wibberley, 1977; Coleman, 1978).

Cross-national comparisons constitute the most challenging task, because definitional problems baulk even larger in this context and because the statistics relate to a greater range of dates. In connection with the World Land Use Survey, however, Stamp (1965) pioneered the compilation of land-use statistics for some twenty European countries, relating to around 1960, an exercise which was updated to around 1965 by Corver and Kippers (1969). Since then a series of major studies has been carried out at Wye College, beginning with a comparative analysis of England and Wales, the USA, the Netherlands and West Germany (Best and Mandale, 1971) and extending out to include the EEC (Best, 1979) and Canada and Sweden (Best, 1981; Hansen, 1981). This work has largely concentrated on 1971 patterns, though an attempt has been made to examine changes over the previous ten years and parts of the analysis have been taken back to 1951 (Hansen, 1981).

The results of these national and international studies allow some observations on land-use patterns and changes which are significant enough not to be undermined by the sorts of data problems outlined below (Figure 2.2). One major conclusion is the predominance of rural land in all the countries covered, in contrast to popular impressions of the more heavily populated countries being largely built up; even in Belgium and the Netherlands, the urban coverage was no more than 15 per cent in 1981. Meanwhile, the area positively recorded in the two major rural categories of agriculture and forestry represented over four-fifths of total land in most countries and fell below two-thirds only in Sweden and Canada, where the statistics are affected by the vast wastes of mountain regions and more northerly latitudes. The proportion of rural land devoted to agriculture as opposed to forestry tends to vary with latitude and overall population density, while the per capita availability of agricultural land draws the expected contrast between the Old and New Worlds. In absolute terms, the EEC's 95 million ha of agricultural land are indeed not much more extensive than Canada's supply and are dwarfed by the USA's figure of 515 million ha.

Figure 2.2: Land-use Composition and Change in Selected Countries: Countries Arranged According to Extent of Urban Coverage

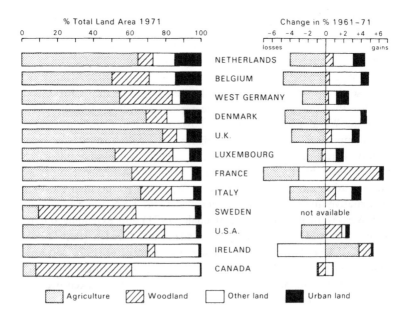

As regards land-use changes over time, these studies indicate a relatively consistent pattern along the lines mentioned earlier, with a general increase in urban area and woodland and a decline in agricultural area, but the review of the problems facing this area of study prompts great caution in interpreting these statistics. There is clearly scope for further improvement in this aspect of rural geography, not merely for updating these kinds of statistics on changing land-use patterns but also including information on a wider range of characteristics, including ownership and interests.

Land Budgeting

Many of the issues relating to the extent and composition of rural land uses revolve implicitly or explicitly round the subject of land budgeting. The setting up of the Scott Committee on land utilisation in rural areas in Britain in 1942 and the recent call for the repetition of this exercise (Gilg, 1978b) stem partly from concern over whether the balance of land uses will be the best one for meeting the nation's needs in the future. Traditionally, the greatest attention has been given to the loss of agricultural land, particularly at the hands of urban development but also in relation to afforestation, mining, water catchment and recreation. Even in the USA, where urban encroachment is not generally reckoned to be a serious problem at the national scale (Hart, 1976), 48 out of the 50 states have adopted forms of land-use controls (Davies and Belden, 1979), while the majority of advanced countries less well endowed with agricultural land have been operating restrictions on urban expansion for some years (Darin-Drabkin, 1977). At the same time, however, problems of agricultural surplus raise doubts about the bases of these policies, as farmland is abandoned to scrub in the USA (Hart, 1968), or recommended for phased reduction in the Mansholt plan for the EEC. Arguments over the existence of a land shortage rage even in such a heavily-populated area as the UK, with Coleman's (1976, 1978) advocation of tighter restrictions on urban expansion being countered by Best's general satisfaction with the present level of control and Boddington's (1973a) anticipation of an agricultural land surplus. Very clearly there is a need for studies which can provide a firmer foundation for policy-making.

A case study of land budgeting for the UK can help to identify the reasons why so much uncertainty still seems to remain. The approach developed by Edwards and Wibberley (1971) is to project the land needs of the major space users (agriculture, forestry and urban development) into the future and thereby see how the total requirements compare with the present overall situation. In these calculations the most detailed treatment is given to agricultural land requirements which are based on a range of assumptions related to population size, per capita incomes, level of national self-sufficiency in food supplies and land productivity. The rate of urban expansion is based on the product of population size and the crude measure of per capita urban land provision, while the future woodland area is estimated with reference to likely trends in financial support for afforestation. The conclusion from that study was that the land budget could be balanced in the year 2000, with none of the land uses having to take over land at the expense of another, as long as no combination of the more pessimistic assumptions was to occur (Edwards and Wibberley, 1971). Adopting the same methodology but a somewhat different set of assumptions, the Centre for Agricultural Strategy (1976) concluded that 'There is little room for manoeuvre . . . Without a more positive approach to land-use planning now, there is a danger of land scarcity in the near future.'

Such studies appear to do little to reduce the level of uncertainty, but at least this approach can provide an indication of the sensitivity of the land budget to the various trends and assumptions. It is the possible range of performance in agricultural yields which affects the budget most dramatically. The level of food self-sufficiency is also important, while the influence of population size and per capita incomes on agricultural land requirements is relatively minor. Rising world pressures on timber supplies could lead to faster afforestation as foreseen more recently by the Forestry Commission (1977), Fairgrieve (1979) and the Centre for Agricultural Strategy (1980), but in terms of both the extent and quality of the farmland affected, the extreme range of 70,000ha to 575,000ha for agricultural-urban land conversions for the period 1975–2000 is the more significant. To some extent these uncertainties arise through an inability to forecast accurately the factors on which the land-use trends are based, such as the demand for timber, the country's trading position and relationship to the EEC, technological developments relating to farmland yields and synthetic foods, and the effect of the changing cost of energy on agricultural production. Some of the uncertainties, however, can also be attributed to the poor state of knowledge concerning the processes of land-use change, particularly the factors affecting the size of the urban land-take. As Coppock (1974) has pointed out, geographers have tended to expend so much energy in collecting land-use statistics that the analysis of land-use patterns and trends has been relatively neglected.

The most common reaction to land-budget uncertainties has been to advocate a virtual embargo on the transfer of agricultural land to other activities. Certainly the Scott Report (Ministry of Works and Planning, 1942) took a strong line in favour of protecting agricultural land from urban encroachment, while the organisers of Land Decade 1980–1990 call for the reclamation of vacant sites within the urban envelope before further agricultural land is alienated (Land Decade Educational Council, 1979; Moss, 1980). Others, however, believe that the fact that food is a basic human need should not rule out the possibility of alternative ways of achieving the same goal. Anderson (1975) has outlined the major implications of substantially increasing agricultural production for both country and town and, on related economic grounds, Ritson (1980) concludes that the UK should aim for only enough self-sufficiency to meet the minimum nutritional standards needed to safeguard national security. Then again, Mellanby (1975) points out that a dietary shift away from livestock products in favour of vegetables and health foods could substantially increase the carrying capacity of Britain's farmland. Taking a wider perspective, Dennison (1942) was one of the first to stress the importance of weighing up the merits of agriculture and urban development side by side, a theme developed more fully by Ward (1957) and Wibberley (1959). Peters (1970) outlined the techniques available for weighing up the costs and benefits of alternative strategies, whether the farmland is being considered for urban development or forestry. More recent accounts have been provided by Boddington (1973a,b) and Whitby and Willis (1981). The techniques, however, often seem to be running ahead of the basic knowledge. It is generally agreed that, in theory, some optimal balance can be reached between more expensive urban living conditions depending on the rate of urban land-take and the density of new urban development (Champion, 1978), but in fact the research on the variation of urban construction and operating costs with changing density and form is only poorly developed despite the pioneer work of Stone (1973).

The Urban Fringe

Irrespective of whether or not the land budget can be balanced at national level, conflicts between urban and rural interests frequently arise at a more localised scale (Gregor, 1957; Berry and Plant, 1978). This is particularly the case in the so-called 'urban fringe' (also known as the 'rural-urban fringe' or 'rurban fringe'), the rather poorly-defined zone of mixed land uses found round urbanised areas (Thomas, 1974). This term most commonly refers to the active margin of urban encroachment extending one or two miles from the edge of the main built-up area, where the continuation of farming is jeopardised by fragmentation of holdings and rising expectations of sale for development. Similar types of interference can be experienced less intensively over a much broader commuting zone, while some types of problem such as trespass may be limited in their occurrence to fields immediately abutting on to built-up land or roads (Munton, 1974). Though reference to the literature on the problems encountered in the urban fringe indicates a fair measure of disagreement over the approach which should be adopted towards urban pressures, it is possible to detect some signs of a shift away from a high level of urban containment towards the encouragement of mixed land uses. In sympathy with this trend, the emphasis has also moved away from pure planning solutions involving zoning and negative controls towards more positive strategies of land management.

The characteristics and problems of the urban fringe are well documented (Johnson, 1974). The dominant feature is the penetration of urban influences into traditionally rural areas, resulting in changes in land use and social structure (Wills, 1945; Golledge, 1960). As land values rise with the increased possibility of development, farmers respond either by gearing down their operations in expectation of quitting (Sinclair, 1967) or by moving into more intensive factory farming (Higbee, 1967; Moran, 1979). With the former leading to lower maintenance standards and the eventual abandonment of farmland and the latter producing a more built-up farming landscape, the urban fringe is characterised by a much wider variety of land-use patterns than the formerly rural area, generally involving a less satisfactory use of land in terms of agricultural efficiency (Coughlin, 1980). Operating problems are compounded by the disruption caused by trampling, vandalism, dogs and pollution emanating directly from the built-up areas or spreading along easy lines of access such as roads and footpaths (Wibberley, 1959). These various problems have a particularly severe impact on agriculture in countries where free market forces produce a widespread scatter of building in the early stages of an area's development, as some farmers sell out before others (Bryant and Russwurm, 1979); but even in Britain where the pattern of development has been subject to strict regulation by local planning authorities for over thirty years, these problems are still significant (ADAS, 1974; Coleman, 1976; Low, 1973; Munton, 1974). Concern is often intensified by the facts that urban development tends to seek out the flat, well-drained land that is most valuable for agriculture and that the main cities are often located in the most productive agricultural regions of the country (Chisholm, 1962, 1979; Gierman, 1977; Manning and McCuaig, 1977; Neimanis, 1979).

The most common public response to these problems is to control the extent and distribution of rural-urban land conversion through zoning schemes and development plans. In Britain the urban containment strategy adopted in the 1940s as a result of the recommendations of the Barlow and Scott reports has attempted to impose very strict limits on the lateral extension of the major cities, while at the same time channelling surplus development pressures to other sites located at some distance from the main centres and separated from the latter by a protected zone of farmland (Hall, 1976). This policy was reinforced in 1955 by a government circular directing local planning authorities to designate such zones as 'green belts', where controls on development would be enforced even more strongly than in other rural areas (Cullingworth, 1979). In the final analysis this urban containment approach aims at the elimination of the urban fringe, leaving the land as either 'townscape' or 'farmscape' with a clear demarcation between the two (Coleman, 1969, 1977).

The British approach to the urban fringe has been criticised on both theoretical and practical grounds. In the first place, the concept of an extensive green belt zone separating town from country is contrary to the original notions of Ebenezer Howard who put forward his garden-city and social-city models as a method for capitalising on the advantages of both town and country. Secondly, the green belt has been used more as a town-planning device than as a method for promoting the better use of rural land. The aims set out in the 1955 circular on British green belts concentrate on the prevention of coalescence of major cities, the elimination of low-density sprawl which is expensive to service, and the enhancement of the physical setting of cities with a valuable architectural heritage. The protection of agricultural land for its own sake was given relatively low priority, while it was only later that the green belts were officially recognised as holding considerable potential for recreational provision (Thomas, 1970). In fact studies have shown that restrictions on development have not been adequate to prevent the rundown of farming activities in the urban fringe. It is not always clear, however, whether the cause is a deterioration of the operating environment despite these controls, or whether changes in the technology and economics of the farming industry have made the previous conditions less viable (Davidson and Wibberley, 1977; Coleman, 1976).

The alternative which has been receiving increasing attention over the last few years is a more purposeful approach to the urban fringe, recognising it as a particular type of area which cannot be zoned away even under conditions of zero urban growth and which offers several opportunities for producing a more efficient and attractive environment. Davidson (1974a), in particular, has stressed its advantages for recreation, emphasising its accessibility to the urban population and the ease with which waste and derelict land can be transformed into recreation space. Rettig (1976) has drawn on the example of the 'greenways' and 'blueways' strategy in Stoke on Trent, the Flevohof educational/ recreational farm in the Netherlands, and farmland consolidation schemes in France and the Netherlands in his attempt to show how recreation, forestry, agriculture and some limited building can be combined efficiently and attractively within an urban fringe setting.

The main emphasis in this alternative approach, however, lies not so much with the preparation of physical plans for the guidance of private investment as with an active government involvement in the day-to-day management of the area. Though public acquisition of land is advocated by some, particularly for recreational developments (McNab *et al.*, 1982), most schemes involve dealing with the existing interests in the urban fringe, including farmers and private developers. The experiment at Milton Keynes (Boddington, 1970; Harrison *et al.*, 1971) demonstrates the extent to which effective management can minimise the negative aspects of the transitional period in an area undergoing extensive development, while the Countryside Commission has undertaken pioneer work in a number of highly pressurised areas such as the Bollin Valley (Hall, 1976). The need for such positive planning and management of urban fringe areas may well be increased as a result of a shift in political attitudes away from conservation towards development during economic recession (O'Riordan, 1980) and because of the less secure status of green belts under the structure planning system (Elson, 1981).

Marginal Fringe

The term marginal fringe is used here to denote the zone which, in Coleman's (1969) terms, separates 'farmscape' from 'wildscape' (Figure 2.1) and which is therefore characterised by an intermixture of productive rural land and the wilderness areas where man has made virtually no impact. In fact, the term 'wildscape' should probably be confined to those areas of wilderness which are rarely visited, such as the less accessible mountains, the tundra and the special areas where all forms of exploitation have been prohibited by law. As such, in many countries the marginal fringe constitutes a much more substantial area than the urban fringe, embracing broad swathes of upland in middle latitudes and wide lowland zones in higher latitudes, as well as other areas affected by soil and rainfall constraints. Lying at the extensive margin of production and land rents, it is an area where one might expect few development pressures and little conflict between land users, but in fact a number of issues have led to the evolution of a planning response on lines not so different from the changing attitudes towards the urban fringe and green belt areas.

The classic problem of these areas is their marginality in economic terms (Ashton and Long, 1972; Tranter, 1978). As technology has advanced, it has appeared to favour the better land more than proportionately, thus reducing profit margins on poorer land. In Britain the so-called 'moorland edge' is now considerably lower than at its peak in the 1880s (Parry, 1977), though national output is currently much greater than then; while in recent years Canada has witnessed a general retreat of the edge of the improved farmland area (Beattie *et al.*, 1981). Such farm abandonment can involve a 'domino' effect as the expansion of scrub threatens to infest adjacent holdings retained in productive use and as the decline in population leads to a withdrawal of essential services (Bracey, 1970). Many countries have attempted to arrest this process for various reasons, including the maintenance of the distinctive contribution which the uplands make to livestock rearing (Davidson and Wibberley, 1956), the need to manage these areas in such a way as to provide recreational opportunities (Coppock and Duffield, 1975), and the desirability of keeping a minimum standard of services for use by tourists (Whitby and Willis, 1981).

The solution to the problem of economic viability normally takes the form of public subsidies to the main production sectors of farming and forestry (Gilg, 1978a), but this in its turn has been found to generate further problems for land-use planning. In the first place, farming and forestry themselves tend to compete for the same more fertile land. Even where forestry is regulated to the poorer land, as is the general policy in Britain, afforestation schemes tend to disrupt farming operations by cutting off the upland grazing areas from the lower-lying wintering quarters (Natural Resources Technical Committee, 1957). Secondly, the intensification of agricultural activities has not always fitted in so amicably as more extensive grazing with the other traditional interests such as hunting and shooting, water gathering and military training (Land Use Study Group, 1966). The most serious problems, however, have arisen as a result of the growth in recreational and tourist activities and the increased appreciation of landscape quality and nature conservation (Advisory Panel on the Highlands and Islands, 1964). The open aspect and distinctive ecology of the British moorlands (highly prized despite their relatively recent creation by man's own hand) has been threatened by afforestation (Crowe, 1966) and by ploughing (Porchester, 1977); while the interests in water catchment, military training and mining have disturbed the landscape and have tended to limit access to these areas (Gibbs and Whitby, 1975). Meanwhile, recreation has brought its own threats both to the landscale, in the form of pressure for related development including second homes, and to the often fine ecological balance through trampling and litter (Coppock, 1977b, 1980).

Moves towards more integrated planning and management have generally been recommended as a solution to such conflicts, just as in the urban fringe, but in the marginal fringe developments have taken place more slowly. Though these areas fall within the jurisdiction of the local planning authorities, the latter have had few powers at their disposal relating to rural land apart from its protection from urban development. The real powers continue to be wielded by the various agencies with development interests involving the single use of sites for their land, water and mineral resources. Some of the bodies, particularly those concerned with the woodland area (Forestry Commission, 1971; Carroll, 1978), are increasingly operating within a wider perspective. The earliest steps towards reconciling these interests in an integrated framework took place in the National Parks — owing to the value placed on their aesthetic quality and the recreational pressures therefore arising (Cherry, 1976). The ten National Parks set up in England and Wales after the National Parks and Access to the Countryside Act of 1949 are administered by single authorities under the overall guidance of the Countryside Commission, unlike the concept of Areas of Outstanding Natural Beauty which merely represent designation of special protection in a similar manner to the green belt principle (Gilg, 1978a). A comprehensive approach was also the main aim behind the Highlands and Islands Development Board, set up in 1965 (HIDE, 1981), and the concept of the Rural Development Board which was used briefly in the late 1960s (North Riding Pennines Study Working Group, 1975), though in these last two examples greater emphasis was placed on economic development than in the National Park context.

The great test of these initiatives is to strike an acceptable balance between development and conservation (Wibberley, 1976). This challenge can partly be tackled at a strategic level, but in the final analysis it involves dealing with individual sites and actors. Some success has been achieved in pioneer schemes involving the broad zonation of these marginal areas into sections where particular combinations of uses should be given priority and in the related manipulation of road, parking and recreational provision. In this way, for instance, both the plan for Dartmoor (Devon County Council, 1973) and the road management scheme in the Peak District National Park (Countryside Commission, 1972) aim to channel visitor pressures to areas where least disruption will be caused to other interests, particularly agriculture and wildlife conservation. At the more local scale, attempts have been made to encourage landowners to be more receptive to the idea of recreational activity on their land. As outlined by Blacksell and Gilg (1981), the 1949 Act made provision both for access agreements and management agreements, but these schemes are generally voluntary and offer no financial compensation to the farmer for restricting the nature of his operations. Nevertheless the Upland Management Experiment, begun in 1969 in part of the Lake District National Park, seems to have been successful enough to encourage its continuation and extension (Countryside Commission, 1976).

In the longer term, however, the issue of managing public relations is less significant than the problems of securing the best use of the land and of maintaining the inherent soil qualities in a marginal and fragile natural environment. This challenge is made no easier by the fact that the marginal fringe responds dynamically to wider economic, social and political forces and, like the urban fringe, is not something that can be zoned out of existence. The assessment of land capability is therefore fundamental in planning for these areas and has taken a number of forms. Statham (1972), for instance, experiments with capability analysis and Parry (1977) uses the analysis of past land-use changes in the moorland fringe as a framework for guiding policy. From a more theoretical stand, Selman (1978) reviews four main approaches to reconciling interests and attempting to achieve a degree of multiple use, revolving around the concepts of economic viability, resource-base capacity, individual decision-making behaviour and competing claimants. Recognition of these four groups of planning techniques merely serves to underline the complexity of the interrelationships facing those involved in land management on the physical margin, a subject which is followed up in greater depth by Thomas and Coppock (1980).

Farming and Conservation

As a concept in land-use management, conservation dates from the end of the nineteenth century, with the National Parks Movement in the USA and the establishment of voluntary groups concerned with the protection of rare species and their habitats (O'Riordan, 1971). Only in the last few years, however, has this interest in land taken on such importance that it now ranks alongside agriculture, forestry, recreation and other uses in many land-use conflicts (Davidson, 1974b). Moreover, conservation matters are also distinctive because they relate as much to the main agricultural area as to the marginal contexts discussed in the last two sections (Davidson and Lloyd, 1977). This recent growth in concern over the farmscape can be attributed jointly to the general increase in environmental awareness which took place in the 1960s and to changes in the nature of agricultural activities (Wibberley, 1976).

   One line of research on this topic has documented the nature and scale of the threats posed by a strong and evolving farming sector. The most conspicuous change has been the opening up of the landscape by the grubbing out of hedges and replacement of ditches by underground drains. Hooper and Holdgate (1969) estimated that around 8,000km of hedgerows were being lost annually in the early 1960s in England and Wales, with deleterious consequences for both the landscape (Westmacott and Worthington, 1974) and wildlife (Moore, 1969). The increased use of chemicals in agriculture has also taken its toll of wildlife, with some pesticides like DDT having the most dramatic effects until banned in many countries. Herbicides are much more widely used and have far-reaching implications (Nature Conservancy Council, 1977). Inorganic fertilisers, together with nitrogenous slurry from intensive livestock units, have produced fundamental changes in wetland ecosystems, which have also been eroded by drainage schemes designed to improve the agricultural value of the land, as in the East Anglian Broadlands (Moss and O'Riordan, 1979).

Since Rachel Carson (1962) first drew widespread attention to the harmful effects of new agricultural practices, the conservation movement has developed into a considerable lobby in land management. Strutt (1978) acknowledged that 'There is evident concern about the harmful effect of many current farming practices upon both landscape and nature conservation, coupled with a widespread feeling that agriculture can no longer be accounted the prime architect of conservation nor farmers accepted as the "natural custodians of the countryside." ' Even before this, it had been noticed that farming was in danger of becoming its own worst enemy, as the shift towards heavier machinery and continuous cereal cultivation produced signs of soil breakdown on clay lands (Agricultural Advisory Council, 1970). Moreover, the recent growth in interest has been associated with a shift in attitudes among conservationists themselves away from a protective and rather negative approach into an attacking position advocating a rethinking of priorities in land-use planning and management, as evidenced by Mabey (1980), Shoard (1980) and Green (1981).

The general impact of the conservation movement on rural land use has, however, tended to be weakened by internal divisions in interests and policy attitudes. The clearest split has developed between two major objectives: (a) the protection and provision of access to natural landscapes for their scenic beauty for informal recreation; and (b) the protection of wildlife for research and education (Mabey, 1980). This dichotomy has become particularly marked in Britain, where it is reflected in the division of responsibility for rural conservation between the Countryside Commission and the Nature Conservancy Council. The former is concerned primarily with outdoor recreational provision, landscape planning and land management problems in sensitive areas like the urban fringe and the uplands, while the Nature Conservancy Council is responsible for establishing and maintaining a national set of protected areas intended to represent the range of British ecosystems. Though most other countries have managed to avoid this institutionalised form of division (Green, 1981), even those which are well endowed with land in relation to their population size contain examples of sites where recreational pressures pose a threat to the survival of primeval natural systems and classic ecological structures.

A related issue concerns the spatial extent over which conservation should be imposed. Green (1981) identifies four kinds of amenity conservation policy, namely the protection of species, promotion of education, safeguarding of key areas and ensuring better management of all areas. In giving greater emphasis to ecological considerations, Green favours a new approach to the countryside in which part of the land – certainly a more extensive proportion than the present Nature Reserves and sites of Special Scientific Interest in Britain – should be acquired for the sole purpose of amenity conservation. He outlines the main examples of the types of ecosystems which should be conserved in this way and presents guidance for the selection of sites and for their management in ecosystem, visitor and estate terms. Others such as Mabey (1980) and Shoard (1980), however, advocate a much wider approach to countryside conservation, with a general curb on agricultural intensification. Here further questions arise concerning whether or not farmers should be compensated for any restrictions imposed on their activities, a topic which was hotly disputed during the progress of the Countryside and Wildlife Bill to the British Statute books in 1981.

These arguments over the treatment of conservation and recreation in the farmscape therefore revolve principally around the feasibility of multiple land use. The Countryside Review Committee (1976) reckons that Britain 'does not have room for the luxury of a single use system of designation' and advocates the search for a 'consensus approach' towards resolving rural land-management issues. Meanwhile, however, Wye College (1980) emphasises how divergent and conflicting some of the interests in rural land are in the lowland agricultural areas. There lies a world of difference between voluntary management agreements between rural interests and the imposition on farming and the countryside of the type of planning controls that were introduced over thirty years ago for restricting urban development. It is therefore encouraging to see the degree of compatibility between land uses which some recent research has discovered (see, for instance, Figure 2.3), though the position of agriculture remains the most serious stumbling block.

Conclusion

The last few years have seen some substantial changes of emphasis and approach in the study of rural land use. These have arisen partly from the pattern of real events and partly from developments on the academic front. The lower emphasis being placed on the loss of agricultural land to urban development reflects the slowdown in land losses

Figure 2.3: A Compatibility Matrix of Competing Rural Activities

Key
. compatible or rarely competing
/ incompatible but conflict rare or restricted
X conflicting

Group	Activity	arable cultivation	ley grazing	unenclosed grazing	softwood forestry	hardwood forestry	coppice production	mineral extraction	water supply	drainage, canalisation, sea wall construction	MOD training	sailing	water-skiing	fishing	shooting	riding	bird-watching	rambling	picnicking	wildlife protection	maintenance ecological sites	maintenance archaeological sites	maintenance geological and physical features	landscape protection	air quality	water quality
exploitative	arable cultivation																									
exploitative	ley grazing	/																								
exploitative	unenclosed grazing	X	X																							
exploitative	softwood forestry	/	/	X																						
exploitative	hardwood forestry	/	/	/	/																					
exploitative	coppice production	X	X	.	X	/																				
exploitative	mineral extraction	X	X	/	.	.	.																			
exploitative	water supply	X	X	.	.	.	.	/																		
exploitative	drainage, canalisation, sea wall construction	.	.	.	.	.	.	X	X																	
exploitative	MOD training	/	/	.	.	.	.	.	.	.																
recreational	sailing	.	.	.	.	.	.	.	.	.	/															
recreational	water-skiing	.	.	.	.	.	.	.	.	.	.	X														
recreational	fishing	X	X	.	.	.	.	.	.	X	.	X	X													
recreational	shooting	X	/	.	.	.	.	.	.	X	.	/	X	.												
recreational	riding	X	/	.	.	.	.	.	.	.	/	.	.	.	.											
recreational	bird-watching	X	/	.	.	.	.	.	.	X	/	X	X	.	X	.										
recreational	rambling	X	/	.	.	.	.	.	.	.	X	.	.	.	.	.	.									
recreational	picnicking	X	/	.	.	.	.	.	.	.	X	.	.	.	.	.	.	.								
protective	wildlife protection	X	/	.	.	.	.	.	.	X	.	X	X	/	/	/	/	/	/							
protective	maintenance ecological sites	X	X	.	X	/	.	X	/	X	.	/	/	.	.	/	.	/	/	.						
protective	maintenance archaeological sites	X	/	.	X	X	.	X	/	.	.	.	.	.	.	.	.	/	.	.	.					
protective	maintenance geological and physical features	.	.	.	.	.	.	X	.	X	.	.	.	.	.	.	.	.	.	.	.	.				
protective	landscape protection	X	/	.	/	.	.	/	/	X	.	.	.	.	.	/	.	/	.	.	.	.	.			
protective	air quality	X	/	.	.	.	.	.	.	.	.	.	.	.	.	.	.	.	.	.	.	.	.	.		
protective	water quality	X	/	.	.	.	.	/	.	/	.	/	/	.	.	.	.	.	.	.	.	.	.	.	.	
protective	rural life	/	/	.	.	.	.	.	X	/	/	.	X	.	.	.	.	X	/	.	.	.	.	/	.	.

due to the effects of economic recession on the construction industry and the tendency towards overproduction of agricultural commodities on the world market. The recognition of conflicts between rural land-using interests stems partly from the real changes in the nature of agricultural activites and the upsurge of public concern over environmental matters, but also coincides with the acceptance of a conflict view of society in urban plánning studies. The shift away from plans towards management has perhaps not been so marked in the countryside as in the towns, since in the former land-use planning has adopted a relatively low profile in the face of the strong sectoral interests (agriculture, forestry, water), which are hampering the trend towards a more integrated approach.

The scope for further research on the rural land-use topics covered in this review clearly remains extremely wide. As regards land budgeting, a major need is for studies which can take a balanced view of the economic and social costs and benefits of alternative levels of rural-urban land conversion and thereby test the polarised views of optimists and pessimists. Particularly important is to discover the robustness of the outcomes under different conditions, and to evaluate strategies which offer alternative levels of flexibility in future land-use choices. The question of the need for multiple land use represents another area where costs and benefits in relation both to society in general and to the individual groups involved, need to be identified more clearly. The controversy over the best approach to wildlife protection and ecological conservation is merely the most recent of a series of issues concerning the compatibility of land uses.

The arguments over the viability of a consensus approach to the resolution of land-use conflicts in the countryside require the monitoring of attempts to secure a more integrated approach to countryside planning and, in particular, of the management experiments adopted in the more problematic areas. But, in the absence of government action, it is likely that the main time and effort of academics will continue to be focused on the mere collection of data aimed at keeping an up-to-date record of actual changes taking place in the way in which the countryside's land resources are being used.

References

ADAS (1974), *Agriculture in the Urban Fringe*, Agricultural Development and Advisory Service, London, Technical Report 30

Advisory Panel on the Highlands and Islands (1964), *Land Use in the Highlands and Islands* (Department of Agriculture and Fisheries for Scotland, Edinburgh)

Agricultural Advisory Council (1970), *Modern Farming and the Soil* (HMSO, London)

Anderson, M.A. (1975), 'Land Planning Implications of Increased Food Supplies', *The Planner*, 61, 382–3

Anderson, M.A. (1977), 'A Comparison of Figures for the Land-use Structure of England and Wales in the 1960s', *Area*, 9, 43–5

Ashton, J. and Long, W.H. (eds.) (1972), *The Remoter Rural Areas of Britain* (Oliver and Boyd, Edinburgh)

Beattie, K.G., Bond, W.K. and Manning, E.W. (1981), *The Agricultural Use of Marginal Lands: A Review and Bibliography*, Working Paper 13, Lands Directorate Canada, Ottawa

Berry, D. and Plant, T. (1978), 'Retaining Agricultural Activities under Urban Pressures: a Review of Land Use Conflicts and Policies', *Policy Sciences*, 9, 153–78

Best, R.H. (1959), *The Major Land Uses of Great Britain*, Studies in Rural Land Use 4 (Wye College, University of London)

Best, R.H. (1968), 'Competition for Land between Rural and Urban Uses' in *Land Use and Resources: Studies in Applied Geography*, Institute of British Geographers, London, Spec. Publ. 1, 89–100

Best, R.H. (1976), 'The Extent and Growth of Urban Land', *The Planner*, 62, 8–11

Best, R.H. (1977), 'Agricultural Land Loss – Myth or Reality?', *The Planner*, 63,

15-66

Best, R.H. (1978), 'Myth and Reality in the Growth of Urban Land' in Rogers, A.W. (ed.) (1978)

Best, R.H. (1979), 'Land-use Structure and Change in the E.E.C.', *Town Planning Review*, 50, 395-411

Best, R.H. (1981), *Land Use and Living Space* (Methuen, London)

Best, R.H. and Coppock, J.T. (1962), *The Changing Use of Land in Britain* (Faber and Faber, London)

Best, R.H. and Mandale, M. (1971), *Competing Demands for Land in Technologically Advanced Countries*, Wye, SSRC Research Project, Final Report

Best, R.H. and Rogers, A.W. (1973), *The Urban Countryside* (Faber and Faber, London)

Blacksell, M. and Gilg, A.W. (1981), *The Countryside: Planning and Change* (Allen and Unwin, London)

Boddington, M.A.B. (1970), *Agriculture in Milton Keynes*, Technical Supplement 9, Plan for Milton Keynes, Milton Keynes Development Corporation

Boddington, M.A.B. (1973a), 'The Evaluation of Agriculture in Land Planning Decisions', *Journal Agricultural Economics*, 24, 37-50

Boddington, M.A.B. (1973b), 'A Food Factory', *Built Environments*, 2, 443-5

Bracey, H.E. (1970), *People and the Countryside* (Routledge and Kegan Paul, London)

Bryant, C.R. and Russwurm, L.H. (1979), 'The Impact of Non-farm Development on Agriculture: a Synthesis', *Plan Canada*, 19, 122-39

Carroll, M.R. (1978), *Multiple Use of Woodlands* (Department of Land Economy, University of Cambridge)

Carson, R. (1962), *Silent Spring* (Hamilton, London)

Centre for Agricultural Strategy (1976), *Land for Agriculture*, C.A.S. Report 1 (University of Reading)

Centre for Agricultural Strategy (1980), *Strategy for the U.K. Forest Industry* (CAS, University of Reading)

Champion, A.G. (1974), 'Competition for Agricultural Land' in Edwards, A.M. and Rogers, A.W. (eds.), *Agricultural Resources* (Faber and Faber, London), pp. 213-44

Champion, A.G. (1975), *An Estimate of the Changing Extent and Distribution of Urban Land in England and Wales 1950-70*, Research Paper 10, Centre for Environmental Studies, London

Champion, A.G. (1978), 'Issues over land' in Davies, R. and Hall, P. (eds.), *Issues in Urban Society* (Penguin, Harmondsworth), pp. 21-52

Cherry, G.E. (1976), *National Parks and Recreation in the Countryside* (HMSO, London)

Chisholm, M.D.I. (1962, 1979), *Rural Settlement and Land Use* (Hutchinson, London)

Clawson, M. (1972), *America's Land and its Uses* (Johns Hopkins Press, Baltimore)

Clawson, M., Held, R.B. and Stoddart, C.H. (1960), *Land for the Future* (Johns Hopkins Press, Baltimore)

Clawson, M. and Stewart, C.L. (1965), *Land Use Information. A Critical Survey of U.S. Statistics Including Possibilities for Greater Uniformity* (Johns Hopkins Press, Baltimore)

Cloke, P.J. (1980), 'New Emphases for Applied Rural Geography', *Progress in Human Geography*, 4, 181-217

Coleman, A.M. (1961), 'The Second Land Use Survey: Progress and Prospect', *Geographical Journal*, 127, 168-86

Coleman, A.M. (1969), 'A Geographical Model for Land-use Analysis', *Geography*,

54, 43–55

Coleman, A.M. (1976), 'Is Planning Really Necessary?', *Geographical Journal*, 142, 411–37

Coleman, A.M. (1977), 'Land Use Planning: Success or Failure?', *Architects' Journal*, 165, 94–134

Coleman, A.M. (1978), 'Agricultural Land Losses: the Evidence from Maps' in Rogers, A.W. (ed.) (1978)

Coppock, J.T. (1968), 'Changes in Rural Land Use in Great Britain' in *Land Use and Resources: Studies in Applied Geography*, Institute of British Geographers, London, Spec. Publ. 1, 111–25

Coppock, J.T. (1974), 'Geography and Public Policy: Challenges, Opportunities and Implications', *Transactions of the Institute British Geographers*, 63, 1–16

Coppock, J.T. (1977a), 'The Challenge of Change: Problems of Rural Land Use in Great Britain', *Geography*, 62, 75–86

Coppock, J.T. (ed.) (1977b), *Second Homes: Curse or Blessing?* (Pergamon, Oxford)

Coppock, J.T. (1980), 'Conflict in the Cairngorms', *Geographical Magazine*, 52, 417–33

Coppock, J.T. and Duffield, B.S. (1975), *Recreation in the Countryside* (Macmillan, London)

Coppock, J.T. and Gebbett, L.F. (1978), 'Land Use and Town and Country Planning' in Maunder, W.F. (ed.), *Reviews of U.K. Statistical Sources*, vol. 8 (Pergamon, Oxford)

Corver, H. and Kippers, M. (1969), *Changing European Land Use Patterns* (United Nations, Food and Agriculture Organisation, Rome)

Coughlin, R.E. (1980), 'Farming on the Urban Fringe', *Environment*, 22, 33–9

Countryside Commission (1972), *The Goyt Valley Traffic Experiment* (CC, Cheltenham)

Countryside Commission (1976), *The Lake District Upland Management Experiment* (CC, Cheltenham)

Countryside Review Committee (1976), *The Countryside – Problems and Policies* (HMSO, London)

Countryside Review Committee (1978), *Food Production in the Countryside: A Discussion Paper* (HMSO, London)

Crowe, S. (1966), *Forestry in the Landscape* (HMSO, London)

Cullingworth, J.B. (1979), *Town and Country Planning in Britain* (Allen and Unwin, London)

Darin-Drabkin, H. (1977), *Land Policy and Urban Growth* (Pergamon, Oxford)

Davidson, B.R. and Wibberley, G.P. (1956), *The Agricultural Significance of the Hills*, Studies in Rural Land Use 3, Wye College, University of London

Davidson, J. (1974a), 'Recreation and the Urban Fringe', *The Planner*, 60, 889–93

Davidson, J. (1974b), 'A Changing Countryside' in Goldsmith, F.B. and Warren, A. (eds.), *Conservation in Practice* (Wiley, London)

Davidson, J. and Lloyd, R. (1977), *Conservation and Agriculture* (Wiley, London)

Davidson, J. and Wibberley, G.P. (1977), *Planning and the Rural Environment* (Pergamon, Oxford)

Davies, B. and Belden, J. (1979), *A Survey of State Programs to Preserve Farmland* (Council on Environmental Quality, Washington, DC)

Dennison, S.R. (1942), *Minority Report of the Committee on Land Utilisation in Rural Areas*, Cmd. 6378 (HMSO, London)

Devon County Council (1973), *Dartmoor National Park Policy Plan* (Devon County Council)

Edwards, A.M. and Wibberley, G.P. (1971), *An Agricultural Land Budget for*

Britain 1965-2000, Studies in Rural Land Use 10, Wye College, University of London

Elson, M. (1981), 'Structure, Plan Policies for Pressured Rural Areas' in Gilg, A.W. (1981)

Fairgrieve, R. (1979), *A Policy for Forestry* (Conservative Policy Centre, London)

Fordham, R.C. (1974), *Measurement of Urban Land Uses*, Occasional Paper 1, Department of Land Economy, University of Cambridge

Forestry Commission (1971), *Forest Management for Conservation, Landscaping, Access and Sport* (Forestry Commission, Edinburgh)

Forestry Commission (1977), *The Wood Production Outlook in Britain* (Forestry Commission, Edinburgh)

Gibbs, R. and Whitby, M.C. (1975), *Local Authority Expenditure on Access Land* (Agricultural Adjustment Unit, University of Newcastle upon Tyne)

Gierman, D.M. (1977), *Rural to Urban Land Conversion*, Occasional Paper 16, Ministry of Supply and Services Canada, Ottawa

Gilg, A.W. (1978a), *Countryside Planning; The First Three Decades 1945-76* (David and Charles, London)

Gilg, A.W. (1978b), 'Needed – a New "Scott" Inquiry', *Town Planning Review*, 49, 353-70

Gilg, A.W. (ed.) (1980, 1981), *Countryside Planning Yearbook*, vol. 1, vol. 2 (Geo Books, Norwich)

Golledge, R.G. (1960), 'Sydney's Metropolitan Fringe: a Study in Rural-urban Relations', *Australian Geographer*, 7, 243-55

Green, B.H. (1981), *Countryside Conservation* (Allen and Unwin, London)

Gregor, H. (1957), 'Urban Pressures on Californian Land', *Land Economics*, 33, 311-25

Hall, A. (1976), *The Bollin Valley – A Study of Land Management in the Urban Fringe* (Countryside Commission, Cheltenham)

Hall, P. (1974), *Urban and Regional Planning* (Penguin, Harmondsworth)

Hansen, J.A. (1981), 'A Land-use Study of Canada, the U.S. and Britain, c. 1951-71', *Area*, 13, 169-71

Hansen, J.A. (1982), *Land Use in North America and the E.E.C., 1951-71*, Occasional Paper 6, Dept of Environmental Studies and Countryside Planning, Wye College, University of London

Harrison, A. *et al.* (1971), *Milton Keynes Revisited* (Department of Agricultural Economics, University of Reading)

Hart, J.F. (1968), 'Loss and Abandonment of Cleared Farmland in the Eastern United States', *Annals of the Association of American Geographers*, 58, 417-39

Hart, J.F. (1976), 'Urban Encroachment on Rural Areas', *Geographical Review*, 66, 3-17

Heap, D. (1978), 'An Outline of Planning Law', *Estates Gazette*

Higbee, E. (1967), 'Agricultural Land and the Urban Fringe' in Gottmann, J. and Harper, R.A. (eds.), *Metropolis on the Move* (Wiley, New York)

Highlands and Islands Development Board (1981), *Annual Report* (HIDB)

Hooper, M.D. and Holdgate, M.W. (eds.) (1969), *Hedges and Hedgerow Trees*, Monks Wood Experimental Station Symposium 4, Nature Conservancy, London

Johnson, J.H. (ed.) (1974), *Suburban Growth* (Wiley, London)

Land Decade Educational Council (1979), *Land Use Perspectives* (Land Decade Educational Council, London)

Land Use Study Group (1966), *Forestry, Agriculture and the Multiple Use of Land* (HMSO, London)

Low, N. (1973), 'Farming and the Inner Green Belt', *Town Planning Review*, 44, 103-116

Mabey, R. (1980), *The Common Ground* (Hutchinson, London)

McNab, A., Anderson, R., Wager, J. and Cobham, R. (1982), 'Planning the Urban Fringe – a Comprehensive Approach', *The Planner*, 68, 20–1

Manning, E.W. and McCuaig, J.D. (1977), *Agricultural Land and Urban Centres*, Report 11, Canada Land Inventory, Ottawa

Mather, A.S. (1979), 'Land Use Changes in the Highlands and Islands 1946–75', *Scottish Geographical Magazine*, 95, 114–21

Mellanby, K. (1975), *Can Britain Feed Itself?* (Merlin, London)

Ministry of Works and Planning (1942), *Report of the Committee on Land Utilisation in Rural Areas* (Scott Report), Cmd. 6378 (HMSO, London)

Moore, N.W. (1969), 'Experience with Pesticides and the Theory of Conservation', *Biological Conservation*, 1, 201–7

Moran, W. (1979), 'Spatial Patterns of Agriculture on the Urban Periphery: the Auckland Case', *Tijdschrift voor Economische en Sociale Geografie*, 70, 164–74

Moss, B. and O'Riordan, T. (1979), 'Alarm Call for the Broads', *Geographical Magazine*, 52, 47–59

Moss, G. (1980), *Britain's Wasting Acres: Land Use in a Changing Society* (Architectural Press, London)

Munton, R.J.C. (1974), 'Farming on the Urban Fringe' in Johnson, J.H. (ed.) (1974)

Natural Resources (Technical) Committee (1957), *Forestry, Agriculture and Marginal Land* (Zuckerman Report) (HMSO, London)

Nature Conservancy Council (1977), *Nature Conservancy and Agriculture* (Nature Conservancy Council, London)

Neimanis, V.P. (1979), *Canada's Cities and Their Surrounding Land Resource*, Report 15, Canada Land Inventory, Ottawa

North Riding Pennines Study Working Group (1975), *North Riding Pennines Study: Report* (North Riding County Council)

O'Riordan, T. (1971), *Perspective on Resource Management* (Pion, London)

O'Riordan, T. (1980), 'Environmental Issues', *Progress in Human Geography*, 4, 417–32

Parry, M.L. (1977), *A Framework for Land Use Planning in Moorland Areas*, Occasional Paper 4, Department of Geography, University of Birmingham

Peters, G.E. (1970), 'Land Use Studies in Britain: a Review of the Literature with Special Reference to the Applications of Cost-benefit Analysis', *Journal of Agricultural Economics*, 21, 171–214

Pierce, J.T. (1981), 'Conversion of Rural Land to Urban: a Canadian Profile', *Professional Geographer*, 33, 163–73

Porchester, Lord (1977), *A Study of Exmoor* (HMSO, London)

Rettig, S. (1976), 'An Investigation into the Problems of Urban Fringe Agriculture in a Green Belt Situation', *Planning Outlook*, 19, 50–74

Rhind, D. and Hudson, R. (1980), *Land Use* (Methuen, London)

Ritson, C. (1980), *Self-sufficiency and Food Security*, Paper 8, Centre for Agricultural Strategy, University of Reading

Rogers, A.W. (ed.) (1978), *Urban Growth, Farmland Losses and Planning*, Rural Geography Study Group, Institute of British Geographers

Secretary of State for Scotland (1973), *Land Resource Use in Scotland*, Cmnd. 5248 (HMSO, Edinburgh)

Selman, P. (1978), 'Alternative Approaches to the Multiple Use of the Uplands', *Town Planning Review*, 49, 153–74

Shoard, M. (1980), *The Theft of the Countryside* (Temple Smith)

Sinclair, R. (1967), 'Von Thünen and Urban Sprawl', *Annals of the Association American Geographers*, 57, 72–87

Stamp, L.D. (1931), 'The Land Utilisation of Britain', *Geographical Journal*, 78, 40-53

Stamp, L.D. (1965), *Land Use Statistics of the Countries of Europe*, World Land Use Survey, Occasional Papers 3, Geographical Publications

Statham, D.C. (1972), 'Natural Resources in the Uplands: Capability Analysis in the North York Moors', *Journal Royal Town Planning Institute*, 58, 468-78

Stone, P.A. (1973), *The Structure, Size and Costs of Urban Settlements* (Cambridge University Press, Cambridge)

Strutt, N. (1978), *Agriculture and the Countryside* (Advisory Council for Agriculture and Horticulture in England and Wales, London)

Thomas, D. (1970), *London's Green Belt* (Faber and Faber, London)

Thomas, D. (1974), 'The Urban Fringe: Approaches and Attitudes' in Johnson, J.H. (ed.) (1974)

Thomas, M.F. and Coppock, J.T. (eds.) (1980), *Land Assessment in Scotland* (The University Press, Aberdeen)

Tranter, R.B. (ed.) (1978), *The Future of Urban Britain*, Centre for Agricultural Strategy, University of Reading

Ward, J.T. (1957), 'The Siting of Urban Development on Rural Land', *Journal Agricultural Economics*, 12, 451-66

Westmacott, R. and Worthington, T. (1974), *New Agricultural Landscapes* (Countryside Commission, Cheltenham)

Whitby, M.C. and Willis, K.C. (1981), *Rural Resource Development* (Methuen, London)

White, P.E. (1981), 'The Rising Rural Interest Rate', *Progress in Human Geography*, 5, 604-9

Wibberley, G.P. (1959), *Agriculture and Urban Growth* (Michael Joseph, London)

Wibberley, G.P. (1976), 'Rural Resource Development in Britain and Environmental Concern', *Journal Agricultural Economics*, 27, 1-16

Wills, N.R. (1945), 'The Rural-urban Fringe: Some Geographical Characteristics', *Australian Geographer*, 5, 29-35

Wye College (1980), *Comments to the Countryside Review Committee* (Department of Environmental Studies and Countryside Planning, Wye College, University of London)

3 STRUCTURAL CHANGE IN AGRICULTURE

I.R. Bowler

Farm structure — the size and spatial disposition of land holding — must be viewed as a fundamental factor in agricultural production. From it stem economies or diseconomies of scale in the use of labour and machinery; often it controls the availability of capital for investment in existing or new forms of agricultural production; increasingly, economies external to the farm, for example in the marketing of produce or the purchase of farm supplies, are denied to the occupiers of small or fragmented farms; above all farm structure is a major determinant of farm income.

The term 'structural change' is generally applied in two agricultural contexts. One concerns alterations to the spatial arrangement of the fields comprising individual farms, while the second involves changes in the number and size of individual farms. So broad is the topic of structural change, however, that some limits must be imposed for the present analysis. Consequently structural changes under land reform and collectivisation have been set aside. Although in a world perspective they are probably the most significant types of structural change at the present time, they have received exhaustive analysis already from, amongst others, Dorner (1972), Jackson (1971), King (1977) and Symons (1972). Rather, attention is focused on the processes of land consolidation and farm enlargement, and for convenience on the context of developed economies.

Land Fragmentation and Consolidation

The Causes of Land Fragmentation

The causes of land fragmentation can be grouped under three main headings: socio-cultural, economic and physical (Burton and King, 1981). Socio-cultural causes appear to predominate, with inheritance laws playing a particularly influential role (Dovring, 1965; OECD, 1972). Partible inheritance, whereby land is equally subdivided amongst the heirs of a landowner, leads to the successive fragmentation of plots of land and can become problematic within relatively few generations.

46

The Napoleonic Code of inheritance provided the legal basis for this practice in France, while in parts of West Germany it is termed 'gavel-kind' (Mayhew, 1970). A rapidly growing population and lack of alternative sources of employment can also lead to fragmentation. In localised areas where, through religious practice, families tend to be particularly large – for example in the Roman Catholic areas of south-east Netherlands (Vanderpol, 1956) – the problem of fragmentation can be exacerbated. Problems of fragmentation are generally absent in countries such as the United Kingdom where, from the sixteenth century onwards, the Enclosure Movement established a structure of relatively large farms. In addition there has been a long tradition of passing land to the eldest son. In Denmark and Sweden also, through reforms beginning at the end of the eighteenth century, plots of land have been brought together in well-arranged units (OECD, 1972).

There are, however, a number of economic causes of land frag-mentation. The piecemeal reclamation of moorland and the drainage of marshland, for example, can result in a scattered pattern of land ownership, as can the fossilisation of open-field patterns inherited from historic systems of communal farming (Lambert, 1963). Equally the gift, sale or purchase of plots of farm land, perhaps to provide a dowry, raise capital to repay a debt or increase the size of a farm, also result in land fragmentation (Binns, 1950). This last facet has an additional agricultural rationale when, by owning plots of land with different soil and micro-climatic properties, the risks associated with drought, flood, frost and wind exposure are spread. Moreover, where a range of crops is grown, each crop being favoured by particular soil or moisture conditions, an individual may require a variety of land quality for farming success. In a third group of causes lie features such as high or broken relief. The scattered patches of more fertile or level land lead almost inevitably to a fragmented pattern of land ownership.

The Effects of Land Fragmentation

At a low level of economic development land fragmentation has certain advantages. For example, it is an ecologically adaptive strategy reflecting variations in land quality; it acts as a check on the diffusion of crop and animal diseases since a multiplicity of plot ownerships tends to produce a mixed pattern of land use.

The advantages of fragmentation, such as those claimed by Igbozurike (1970) and Johnson (1970) for tropical areas, appear to depend on a number of limiting conditions. Amongst these are the existence of spatial variability in land quality, a weakly developed commercial orientation in agriculture, low costs of transport, a low level of use of capital-intensive farming techniques, and a high cost for the exchange of parcels of land. Once a commercially-oriented agriculture develops, fragmentation increasingly imposes economic costs on farming. For example, the cost of transporting labour to scattered plots of land becomes significant in determining the level of profitability of a farm. Up to one-third of farming time can be used just in travelling to and fro between individual parcels of land. The cost can be measured either directly in the wage paid to a farmer or farm worker while travelling unproductively to a distant plot of land, or indirectly as the opportunity cost of the travel time (Chisholm, 1979). In addition the small size and irregular shape of individual plots of land, together with the difficulty of gaining access across land owned by other farmers, inhibits the use of cost-reducing farm machinery. If animals rather than crops are being farmed, the time-consuming movement of livestock from plot to plot is incurred together with the high cost of fencing or supervision.

Other farming developments are also limited by land fragmentation. These include the improvement of the soil by drainage or ploughing, the rotation of crops to maintain soil fertility, and the provision or control of irrigation. Since commercial farming also leads to more specialised types of agriculture, the rationale for owning plots with different land characteristics is also eroded. Rather more difficult to establish are the 'social-psychological' problems of a farming community, as advanced by Burton and King (1981), which include disputes over access to plots and litigation over contested ownership.

The impact of fragmentation on agriculture is well known (Found, 1971; Chisholm, 1979). Less time is spent by the farmer on his more distant fields especially those involving a journey in excess of an hour or 3-4 kilometres in distance. More labour-extensive land uses tend to be practised with increasing distance from the farmstead and there is a corresponding decline in output per hectare. These phenomena, which appear in both temperate and tropical farming systems (Blaikie, 1971; Jackson, 1972), are often quoted in support of normative models of agricultural location. In some regions more distant plots of land are abandoned altogether, while overall the land is used sub-optimally.

A fragmented pattern of land ownership also fosters a nucleated pattern of farmsteads, there being no advantage in locating on any particular parcel of land. The resulting congestion in farm villages is not conducive to the development of a modern farm layout or to the access of machinery to the farmstead.

The Measurement of Land Fragmentation

Statistics on the degree of land fragmentation are collected from time to time by agricultural census in those countries where it is a problem. Usually the data are published in aggregate form for regions within a country, but there is a lack of comparability between countries in the scope and timing of the censuses. Moreover, field patterns of ownership, and distances separating parcels of land from the farmstead, generally have to be identified on a farm-by-farm basis in the field by government officials or individual researchers. The resulting cadastral survey not only becomes out of date within a few years, but at the time is at best imprecise and incomplete. Problems arise from the prolonged absence of some owners from their lands, while others may be reluctant to disclose the full extent of their land ownership. Many areas to the present day have no up-to-date cadastral maps upon which to base studies of land fragmentation.

Even when field surveys can be completed there are difficulties in generalising the very diverse spatial patterns of plot ownership. Several indices of fragmentation have been developed using parameters such as the average size of individual plots, the number of plots per farm, and the shape of plots. The more elementary indices merely relate the number and area of plots to the total area of a farm (Januszewski, 1968; Simmons, 1964). More recently attempts have been made to measure the spatial dispersion of plot ownership (Bryant, 1975; Igbozurike, 1974), but as yet the actual spatial patterns of fragmentation remain poorly exposed.

Given the problems of measurement, it is not surprising that global estimates of the degree and location of land fragmentation have been published infrequently. However, the problem of fragmentation appears to be greatest in Europe (Vander Meer, 1975), particularly in France, West Germany, Greece, Portugal and Spain (Table 3.1). In the mid-1950s, for example, at least one-third of Europe's farmland was considered to be in need of consolidation (Lambert, 1963). This proportion rose to 40 per cent in France and 60 per cent in Portugal (Dovring, 1965). But more detailed investigations reveal marked regional variations in the severity of fragmentation. Thus the problem is more severe in southern Greece (Table 3.2), northern Portugal and north-western Spain than elsewhere in these countries. In Galicia (north-western Spain), for example, O'Flanagan (1980) found an average of 31.5 plots per holding in the farm structure.

Table 3.1: Fragmentation of Agricultural Land[a]

Country	Average number of plots per farm	Proportion of holdings with over ten plots (%)	Average area of arable land per farm (ha)
Greece	7	37	3.2
Italy	4	6	–
Portugal	6	14	5.1
Spain	14	33	7.1
Turkey	7	16	7.7
West Germany	10	31	–

Note: a. Various dates between 1950 and 1962.
Source: OECD, 1969.

Table 3.2: Structure of Farms by Regions in Greece (1961)

Region	Average number of plots per farm	Average area of plots (ha)	Average area of farms (ha)
Thrace	7.6	0.5	3.7
Macedonia	6.7	0.5	3.2
Thessaly	6.9	0.6	4.2
Epirus	5.5	0.4	2.0
Central Greece	7.3	0.5	3.4
Peloponnese	6.7	0.5	3.5
Ionian islands	6.1	0.3	1.9
Aegean Islands	7.3	0.3	2.2
Crete	10.5	0.3	2.9
Total	7.1	0.4	3.2

Source: OECD, 1969.

Land Consolidation Schemes

In an increasingly commercialised world, the disadvantages of a fragmented pattern of land ownership have led to intensified pressures for plot consolidation. A modern, economically efficient agriculture is desired not only by producers seeking higher farm incomes, but also by governments striving to limit the cost of providing financial support for the farm population. Often the price of that support is an increasingly efficient agriculture.

Land consolidation is concerned with the rearrangement and re-allocation of scattered plots of land. Sometimes a simple, voluntary exchange or purchase of plots can be arranged on the initiative of individual owners (Thompson, 1961; White, 1966). More often, however, government intervention is necessary to speed up what is otherwise an extremely slow process. Thus in many countries government authorities, backed by funds, help to promote the exchange of land between individuals. On a wider scale, accelerated consolidation schemes have been promoted, although little effective legislation has been developed outside Europe (King, 1977). In these schemes government finance, and the expertise of a consolidation authority, is made available once a certain proportion of landowners in a designated area agree to a consolidation plan. In Spain the proportion is 50 per cent of landowners or the owners of at least 75 per cent of the land (Naylon, 1959; OECD, 1969); in France grant aid for 'remembrement' is available once 75 per cent of landowners are in agreement; in Cyprus the proportion is 50 per cent (King, 1980). At a third level, integrated schemes of structural reform link plot consolidation to the comprehensive development of the rural economy including settlement, roads, drainage, irrigation, and even off-farm employment. Thus in West Germany a consolidation authority can take the initiative in carrying out integrated schemes ('Flurbereinigungsverfahern') in certain designated areas under the 1953 Land Consolidation Act (Mayhew, 1970, 1973). Initially these schemes were responsible for consolidating a larger area of land than the parallel accelerated consolidation schemes, although they later fell into disfavour owing to their high cost. In the Netherlands also, under the 1954 Consolidation Act, the restructuring of the whole rural landscape has been looked upon as part of the consolidation package (Lambert, 1963).

The outcome of a consolidation scheme should be the concentration of scattered plots into compact holdings, with each holding allocated as near as possible to the new owner's largest existing area of land or farmstead. New parcels need to be accessible by farm road, and to be as large in size, as compact in shape, and as few in number as can be achieved. Thompson (1961), for example, has described the process of 'remembrement' in one commune in France — Vezelise in Lorraine — by which the number of fields was reduced from 2,643 in 1811 to 358 in 1960 (Figure 3.1).

Most countries with the problem of land fragmentation have set up a consolidation authority. In the Netherlands since 1924, for example, the executive body has been the Government Service for Land and Water Use, whereas in Cyprus a Land Consolidation Authority has operated only since 1969. An authority is needed, in the first instance, to determine those areas in greatest need of consolidation, the existing pattern of land ownership, and the value of the land based on the productivity of the soil. This last task is of critical importance since

Figure 3.1: Pre and Post Land-consolidation Field Boundaries in Vezelise (Lorraine)

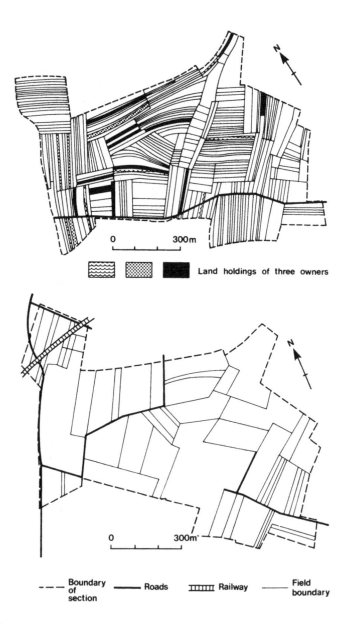

Land holdings of three owners

Boundary of section   Roads   Railway   Field boundary

land of equal value has to be exchanged. Disagreements over valuations are inevitable unless a disinterested authority is available to operate a valuation scheme and arbitrate when disputes arise. Moreover, the authority must formulate a plan for the reallocation of land, including revised field boundaries and new roads, mobilise support for the plan, supervise voting on the proposals, and implement the legal exchange of land ownership (Lambert, 1961).

The Evaluation of Consolidation Schemes

National schemes for plot consolidation have progressed only slowly (Rickard, 1970), while an individual scheme can take up to seven years to reach completion. 'Remembrement' in France took twenty-eight years to consolidate 43 per cent of the land in need of treatment, and that was with a yearly rate of completion superior to any other country (Baker, 1961) (Table 3.3). The broad conclusion to be drawn from international comparisons, as made by Lambert (1963) and OECD (1972), is that it will take most countries about thirty years at their present rate of progress to treat the area suffering from fragmentation. Even this conclusion is misleading for, as Clout (1971) has emphasised for West Germany, a high proportion of the land already consolidated is now in need of consolidation for a second time if contemporary farming requirements are to be met.

The resistance to consolidation schemes centres around the fear of change and the 'psychic' attachment of individuals to particular plots of land (Mayhew, 1971). As King (1977) notes, the implementation of

Table 3.3: Land Consolidation and Related Schemes[a]

Country	Consolidated area (ha): since	Area still to be consolidated (ha)	Average annual rate of consolidation (ha)	Consolidation ratio old:new parcels
Austria	472,000: 1960	850,000	25,000	3.5 : 1
France	6,900,000: 1945	15,000,000	470,000	4 : 1
Japan	1,091,000: n.a.	1,650,000	51,000	3 : 1
Netherlands	360,000: 1924	1,140,000	50,000	2.5 : 1
Spain	3,000,000: n.a.	1,600,000	180,000	7.6 : 1
Switzerland	190,000: 1940	360,000	5,100	4 : 1
West Germany	5,029,000: n.a.	10,100,000	300,000	5 : 1

Note: a. Various dates in the late 1960s.
Source: OECD, 1972, 25.

consolidation schemes is a crusade against inertia and suspicion. In addition the costs involved for the individual are not always met in full by government funds, while shortages of skilled personnel and finance, especially for the large integrated schemes, place limits on the number of schemes that can be put into operation at any one time. In West Germany in 1970, for example, expenses per resettled farm came to about DM 310,000.

The pace of land consolidation has also varied within each country. Areas with fewer but larger landowners, especially if tenanted farms are significant, face fewer problems in developing a consolidation plan compared with areas where small, owner-occupied farms predominate. Owner-occupiers tend to have a strong attachment to particular parcels of land, while large landowners are able to exploit the advantages of a consolidation scheme fully and meet the ancillary costs involved. The age structure of the farm population also influences attitudes towards land consolidation. The younger age structure of farmers in the Old Castile and Leon regions of Spain, for example, has been suggested as the explanation for the greater progress of land consolidation schemes in those areas (OECD, 1969). Finally, land consolidation is physically easier to implement in lowland areas. When these various factors come together, as in the northern and eastern areas of France (Baker, 1961), considerable progress can be made with consolidation schemes. One consequence, however, as stressed by King (1980) for Cyprus, is the growing disparity between areas with and without consolidation schemes in terms of the living standards of the farm population.

Considering the amount of skilled manpower, finance and social disruption involved, relatively few social or economic evaluations of land consolidation schemes have been published (Bunce, 1973). In part this stems from the difficulty of determining the extent to which a consolidation scheme has promoted farming change, and the extent to which change would have occurred in any event. In addition, many benefits are difficult to quantify (OECD, 1972), while there is also the problem of judging the appropriate time-lag before making an evaluation. Is the success or failure of a consolidation scheme to be evaluated after five, ten or twenty years?

Bearing these problems in mind, two broad categories of benefit can be identified. Economic benefits include the introduction of new (or the more efficient use of existing) farm machinery (Thompson, 1961), increased labour productivity from reduced travel time around the land comprising a farm, reduced unit output production costs, and increased farm inputs and crop yields. Butterwick and Neville-Rolfe

(1965) quote increases in farm profits of 20-25 per cent following a consolidation scheme, while an OECD (1972) study for France suggests increases in output per hectare of up to 50 per cent. The extent to which increases of these magnitudes are a function of changed land use rather than more efficient farming is not clear. Certainly greater land-use specialisation and a reduced level of abandoned land accompany consolidation schemes.

The second group of social benefits are less tangible but include the raised status of individual farmers, a more rational outlook on agriculture, and a lower level of friction within the farming community with the removal of disputes over land (Burton and King, 1981). To these benefits can be added the amelioration of social and working conditions of the farm population which reduces the inequality between rural and urban areas in a country, especially when a degree of resettlement has been possible.

Nevertheless, a number of problems have been identified by evaluations of consolidation schemes. The monetary cost is undoubtedly high. One study in West Germany, for example, indicated that the cost of integrated schemes was greater than their yield in terms of improved incomes (OECD, 1972), although Mayhew (1971) stresses that greater savings can be made from grouped rather than isolated resettlements in providing roads, piped water and electricity. Perhaps the major problem, however, lies in the small size of the farm units established by consolidation schemes. From a social point of view plot consolidation without farm enlargement maintains the rural population on the land while increasing the economic viability of each farm. But from an economic viewpoint the amount of land per farm still remains too small to provide a reasonable income. Yet if too much emphasis is placed on farm enlargement, farmers would probably reject a consolidation scheme since many farms would have to disappear. Consequently Jacoby (1971) sees land consolidation not as a limited programme to be accomplished within a given time period, but as a continuing action which is needed for adjusting agriculture to the processes of demographic change and technological development. Taking this view, several countries (Denmark, the Netherlands, Spain, Switzerland) have established minimum plot and farm sizes in their consolidation schemes, while others, such as West Germany under the 1961 Land Transactions Act (Muller, 1964), have legislation to prevent the subsequent subdivision of newly-established land holdings. Inevitably, therefore, land consolidation is drawn into the larger issues of farm enlargement and regional rural development, the latter being necessary to provide off-farm employment for those farmers losing land.

Farm Enlargement

The Causes of Farm Enlargement

Farm enlargement is a spontaneous process in a market economy, but like plot consolidation it acts only slowly on the structure of agricultural holdings. The economic causes of farm enlargement are widely appreciated (Bowler, 1979). Summarising, the demand for agricultural products tends to rise relatively slowly. The demand for food is inelastic in terms of both price and income, while population growth in developed countries is low. Moreover, most of the increase in food expenditure is accounted for by the growing volume of services supplied with the food such as processing, packaging and marketing. The costs of farm production nevertheless continue to rise in line with inflationary pressures elsewhere in the economy. To compound these problems new agricultural technology, while cost reducing, is also output increasing. Faced by relatively static demand, the increased output depresses market prices and farmers are caught in a continuous price-cost squeeze. In the long term, therefore, resources of land, labour and capital are transferred from agriculture into other sectors of the economy. Those producers remaining in agriculture are then able to share resources and returns to production amongst fewer but larger farm units.

Smaller farms are most vulnerable to the price-cost squeeze. The scope for varying the use of capital and labour is limited (HMSO, 1961) and economies of scale are difficult to achieve. Even when small farms, through intensified production, yield a viable income per hectare, there are insufficient hectares to generate a viable total farm income. In addition the poor managerial ability of most occupiers of small farms has been recognised for many years (Carpenter, 1958). The principal feature of farm enlargement, therefore, is the absorption of the land of small farms by larger farms as the occupiers of the former leave agriculture.

The Measurement of Farm Enlargement

Since comprehensive statistics on individual land transactions are seldom published, farm enlargement is generally studied through agricultural census data, despite their limitations. The rate of structural change can be measured either from alterations in the total amount of agricultural land in relation to the total number of farms, or from comparisons of farm size distributions. Unfortunately census data usually refer to agricultural holdings rather than farms. The latter can comprise several holdings for which separate returns may be made. Thus the degree of farm enlargement is under-estimated. When census authorities encourage farmers to make single rather than multiple returns, however, 'paper amalgamations' (Britton, 1977) can cause the census to record a spuriously high rate of increase in the number of large farm units. In addition, census data on farm structure are usually presented as the number of holdings within certain size categories. These categories vary from country to country making international comparisons difficult (European Commission, 1978), while only the net balance between the inflows and outflows affecting farm numbers in each size group are recorded from census to census. Thus a farm-size change is measured only when a holding 'crosses' a size group boundary with the result that the total amount of structural change is understated.

Perhaps more seriously, the scope of many censuses change through time. In the United Kingdom agriculturally insignificant holdings have been excluded from enumeration from time to time (Ministry of Agriculture, Fisheries and Food, 1977) and a quite false impression of the rate of loss of small farms can be created. In the USA eligibility for the census is defined by the value of farm output. The apparent easing up in the rate of decline of commercial farms in recent years, therefore, may merely reflect price rises which have brought many small farms back into the census. Unfortunately such problems are often overlooked in research papers with the result that losses in the total number of farms are exaggerated, while changes in the size of individual farms remain understated. In addition it must be appreciated that area size is an inadequate measure of a farm business. Surrogate measures of business size, such as standard labour inputs (Ashton and Cracknell, 1961), are increasingly used. Some researchers, such as Steeves (1979) and Whitby (1968), have preferred to use data on the turnover of farmers (retirement and recruitment) as more useful measures of structural change in agriculture.

The Process of Farm Enlargement

Central to the process of farm enlargement is the balance between the supply of and the demand for farmland. Land comes on to the market for sale in four main ways (Agricultural Adjustment Unit,

1968): (a) a farmer moves to another farm; (b) a farmer sells some land to raise capital for investment; (c) a farmer leaves agriculture either through death, age retirement or alternative employment; (d) the creation of new farm land by land reclamation. Although exact proportions vary from country to country, and to a lesser extent from region to region, most farm surveys reveal death and age retirement as the principal ways by which land is made available for purchase (50–60 per cent of land), but with age retirement becoming increasingly important (Gasson, 1969). In some areas age and ill-health together account for over one-third of retirements with a further 20 per cent attributable to low farm incomes (Parson, 1977). Positive reasons, such as taking alternative employment, tend to be in a minority.

Land is purchased on the market also in four main ways (Agricultural Adjustment Unit, 1968): (a) a new entrant moves into agriculture; (b) a farmer moves to another farm; (c) a farmer takes on additional land; (d) land being taken out of agriculture for forestry, urban or recreational uses. Most land is purchased by other farmers to enlarge their holdings (40–60 per cent of land). Existing farmers are under strong economic pressures to acquire more land (Britton, 1968) and they are able to spread the cost of their purchases over a larger area compared with new entrants. Existing producers, therefore, are responsible for bidding-up the price of farmland with the result that the main resource transferred out of agriculture is farm labour rather than land or capital.

Land is made available for farm enlargement at a very slow rate. A number of 'survival strategies' can be adopted by farmers not wishing to leave agriculture including the replacement of relatively expensive farm labour by cheaper capital in the form of buildings, plant and machinery. This allows commodities to be produced at lower unit costs. However, since most modern machinery tends to work efficiently on a large area of land, there is still pressure to enlarge the size of farms. Other strategies include increasing the output of the farm business by intensifying production, and gaining economies of scale by cooperating with other farmers in the purchase of farm inputs and the marketing of farm produce. Another option for a farmer is to seek off-farm employment and so become a part-time farmer. In Japan, 88 per cent of farmers are now in this category, while percentages in other countries are: Norway 66, Austria 56, West Germany 55, USA 54 and Switzerland 51 (OECD, 1978).

Despite these 'survival strategies', small farms tend to produce low incomes for their occupiers. The reasons for the low occupational mobility of farm people can be grouped into two categories: reasons for accepting low farm incomes, and reasons which make leaving agriculture difficult (Gasson, 1969). In the first category lie the 'psychic' attractions of farming as a way of life, including the loss of status and apparent admission of failure (Hill, 1962) attached to leaving agriculture. Weerdenburg (1973) for the Netherlands has shown that employment in industry is an unattractive alternative for most farm people, while a British study has emphasised the reluctance of farmers to leave agriculture when a son or daughter shows an inclination to take over the farm business (Agricultural Adjustment Unit, 1968). Government financial support to agriculture also prevents the disparity between farm and non-farm incomes from becoming large enough to overcome the occupational inertia of remaining in farming (OECD, 1965); at the present time this is justified by politicians because farmers leaving agriculture would merely tend to swell the numbers of those unemployed.

In the second category lie the low 'salvage' value of investments on a farm (asset fixity) (Kingma and Samuel, 1977), the lack of educational, industrial and factory skills in the farm population for off-farm employment (OECD, 1965; Guither, 1963; Bultena, 1969), the high average age of farmers, and the direct cash costs of moving to an urban area (OECD, 1975). Gasson (1970), however, prefers to emphasise the imagined rather than the real costs of moving out of farming including the non-pecuniary costs of uncertainty about the future in an urban occupation. There are dangers, nevertheless, in generalising from a few case studies. Steeves (1979), for example, has found a high and not a low rate of mobility both into and out of Canadian agriculture, and there is clearly scope for further research into this topic under contemporary social and economic conditions.

Broadly, a low but accelerating rate of structural change has been found in most developed countries whether measured in farm employment or farm numbers (Table 3.4). Average annual rates of decline increased from under 2 per cent before 1960 to nearly 3 per cent in the late 1960s (Agricultural Adjustment Unit, 1970). Rates of change have varied between countries reflecting the pre-existing size distribution of farms and the degree of income support provided by government policy measures. Within the broad trend the number of farms has declined at a greater rate when the disparity between farm and non-farm incomes has been larger (Hill, 1962; Robson, 1970).

The rate of farm enlargement varies with size of farm. In West

Table 3.4: Average Annual Rates of Change in (A) Agricultural Employment; (B) Farm Numbers

Country	1950-60 (%)		1960-65 (%)		1965-69 (%)	
	A	B	A	B	A	B
Belgium	−2.0	−2.2	−5.1	−3.1	−4.5	−3.8
Denmark	−3.6	−0.6	−2.5	−2.5	−5.3	−4.2
France	−3.6	−2.3	−3.6	−2.9	−3.6	−2.9
Japan	−2.0	−0.2	−3.8	−1.3	−3.7	−1.2
Netherlands	−1.5	−3.1	−3.2	−2.5	−4.3	−3.1
New Zealand	−0.7	−0.3	+ 0.6	−3.1	+ 2.0	−1.7
Sweden	−3.5	−1.9	−7.3	−4.3	−10.3	−5.0
United Kingdom	−2.0	−1.1	−3.8	−1.6	−3.6	−2.2
United States	−3.3	−4.1	−4.5	−2.0	−2.1	−2.8

Note: a. Various time-periods within those indicated.

Source: OECD, 1972.

Germany, for example, farms below 10 hectares in size declined in number at 5 per cent a year between 1967 and 1977, whereas those over 50 hectares increased at the same rate. Indeed in most EEC countries the threshold between decline and increase appears to lie between 20 and 50 hectares (Calmès, 1981), although as Clout (1971) points out the process of farm-size change is so slow that it takes a decade to raise the average size of farm in the EEC by a single hectare.

Regional variations in the rate of farm enlargement can also be detected. The main causes are to be found in the pre-existing farm-size structure of each region, variations in the age structure of the farm population, and the type of farm occupancy. The process of farm enlargement often begins in areas where large farms predominate, such as eastern England (Britton, 1968; Hine and Houston, 1973), but becomes most highly developed where the farm structure brings together small and medium-sized farms. Both Clark (1979) and the Agricultural Adjustment Unit (1970), for example, discovered higher levels of structural change in north east Scotland and northern England respectively where medium-sized farms were able to take over the land of small farms. As Todd (1979) makes clear in a study of Manitoba farming, there is no spatial correspondence between the development of large and the disappearance of small farms. In areas dominated by small farms, the absence of purchasers for farms falling vacant tends to produce lower levels of structural change (Fuller, 1975).

Within an area, however, the supply of land for farm enlargement depends equally on local demographic factors as on farm-size structure

(Hill, 1962; Whitby, 1968). Older farmers have a higher propensity to sell land than younger farmers, while the latter are more likely to enlarge their farms, especially when between 40 and 50 years of age (Gasson, 1969; Smit, 1975). Again it is the spatial juxtaposition of sellers and buyers that creates variation in the rate of farm enlargement and not just the spatial concentration of elderly farmers (MacDougall, 1970).

The role of type of farm occupancy − whether a farm is rented or owner-occupied − is more difficult to establish. Studies in the USA (Hill, 1962) and the United Kingdom (Agricultural Adjustment Unit, 1970), for example, have suggested a higher rate of structural change amongst tenanted than owner-occupied farms. But other research in the United Kingdom (Hine and Houston, 1973) has shown that farms combining owned and rented land have the greatest propensity to increase in size. Similarly for the USA, Van Otten (1980) concludes that since the cost of purchasing new land is so high, renting and leasing are now preferred as the means of increasing farm size. It is rare for there to be marked regional variations in type of occupancy, however, and this factor is more visible at the farm than the regional level of analysis.

A further dimension to the process of farm enlargement has been established by a number of recent studies using farm survey rather than census data. Williams (1972) for Australia, Edwards (1978) for the United Kingdom, and Smith (1975) and Van Otten (1980) for the USA have all shown how farm enlargement leads to a fragmented pattern of farm ownership, especially amongst the largest farms. Although farmers are prepared to pay more for 'near-at-hand' land, they are equally ready to buy farms at a considerable distance to enlarge their businesses (Figure 3.2).

There are three main diseconomies in farm fragmentation: increased production costs through greater distance inputs, decreased land productivity on more distant holdings, and failure to optimise the timing of operations on all farms in the business. Curiously, some agricultural economists have dismissed these problems as 'inconveniences' (Britton, 1968) which are more than compensated for by the economic advantages of larger scale production; others (Johnston, 1972), however, consider that the diseconomies need explicit consideration, but few such studies have been completed. One study for South Australia (Hill and Smith, 1977) found no consistent evidence that more distant land was used for extensive farm enterprises. Indeed the authors identify three advantages in farm fragmentation: the purchase

Figure 3.2: Holdings Comprising More than Two Sections (1969), South Australia

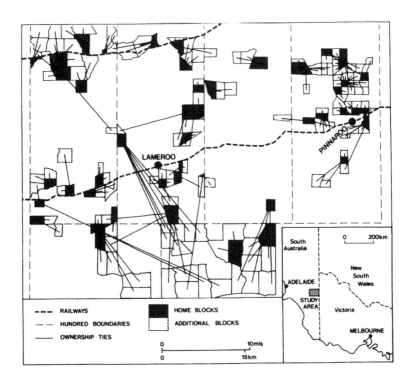

of less expensive land, the spreading of risk associated with weather hazards, and the spreading of operating time for farm labour and machinery. In addition, there is evidence from Australia (Williams, 1970) and the USA (Kollmorgen and Jenks, 1958; Smith, 1975) that in more isolated areas farmers are preferring to live in towns and commute daily to their holdings. Ironically, therefore, a pattern of farm, if not of land fragmentation is being recreated by farm enlargement, and in some areas it is associated with a more nucleated rather than dispersed distribution of the farm population.

Farm Amalgamation Schemes

Governments intervene in the process of farm enlargement for several reasons. One objective is to reduce the number of small, uneconomic holdings which are still farmed on a full-time basis. Their existence makes it difficult for governments to manipulate agricultural product support prices in such a way as to prevent larger farms from receiving windfall profits. In addition governments intervene to speed up farm enlargement, which under market forces is a slow process, and also ensure that the available land is amalgamated with smaller farms whose viability could be enhanced by enlargement.

Structural policies have developed fairly recently in most countries, evolving in the late 1950s out of the failure of price policies to resolve the problem of low farm incomes. Policy measures have been reviewed for developed countries by OECD (1972, 1980), for West European countries by Hirsch and Maunder (1978), and for EEC countries by Rickard (1970) and Fennell (1975). Broadly, three types of policy measure can be identified: retirement pensions, farm amalgamation grants, and retraining schemes.

Retirement pensions, annuities and lump-sum payments are available to farmers retiring voluntarily from agriculture and who meet certain requirements. Usually their farm land must be made available to enlarge neighbouring farms. Thus Sweden has had a retirement compensation scheme since 1967, France has provided termination payments (IVD – *Indemnité Viagère de Départ*) since 1962, and the EEC contributes to all such schemes in member states under Directive 72/160 (Table 3.5).

Credit facilities to enable land to be purchased, and grant aid to assist with the legal and agricultural costs incurred in amalgamating farms are offered to those who remain in agriculture. West Germany, for example, gives low-interest loans for purchasing land under an Individual Farm Investment Promotion Scheme. Generally finance is available only upon application and is paid to those farmers who can show that after enlargement their holdings will be economically viable.

Retraining schemes, especially for the sons of farmers, have been developed most recently, although the case for subsidising occupational and geographical mobility has been recognised for many years (Nalson, 1968; OECD, 1972). Such schemes operate in several countries, including Japan, France and the USA.

The method of administering these various measures differs considerably from country to country. The United Kingdom, for example, has a centralised national authority (Ministry of Agriculture, Fisheries and Food) which administers the schemes, but the sale and purchase of farmland is left to market forces. In Austria, Denmark, Finland, Japan and West Germany, by comparison, the transfer of farmland is subject to public control and approval. Finland's National Board of

Table 3.5: Results of the EEC Directive 72/160, 1975-8

Country	Premiums and annuities (% eligible)	Recipient holdings (% with a development plan)	Area released (% used for development plans)	Size of recipient holdings (ha)		
				< 10	10-20	> 20
				(% each group)		
Belgium	1.9	3.1	5.9	71	25	4
France	1.0	0.7	1.6	40	34	26
Ireland	18.1	66.1	11.0	20	52	28
Luxembourg	—	—	0.3	31	52	17
Netherlands	9.3	1.7	2.1	85	11	4
United Kingdom	6.4	13.3	16.3	7	24	69
West Germany	6.9	26.5	36.3	39	41	20
EEC	4.0	14.8	14.0	—	—	—

Source: European Commission, 1978.

Agriculture, for example, is empowered both to authorise land transactions and to acquire farmland. In 1978 the Board purchased 14,500 hectares and resold 10,000 hectares of land to farmers whose farms needed enlarging.

In other countries government agencies operate at the regional level. Sweden, for example, has had a network of County Agricultural Boards since 1957 (Whitby, 1968). The Boards authorise land transactions but also intervene in the market to buy and sell land. They aim for a more rational pattern of land allocation than that achieved by market forces acting alone. In France the twenty-nine regional authorities are termed *Sociétés d'Aménagement Foncier et d'Etablissement Rural* – SAFERs, and details of their operations have been provided by Butterwick and Neville-Rolfe (1965). The Canadian system is operated at a Provincial level under the 1965 Agricultural and Rural Development Act. The Directorate in each province aims to alleviate rural poverty by enlarging the size of uneconomic farms either by renting or selling land that it has acquired from farmers leaving agriculture. More spatially selective schemes are also practised. A Marginal Lands Committee, for example, operated in South Australia after 1938 to control the pattern of farm abandonment and amalgamation in the drought-affected margins of the wheat-growing areas (Williams, 1976), while a Marginal Dairy Farms Reconstruction Scheme has been applied to dairy farming areas elsewhere in Australia from 1970 to 1974 (Threlfall, 1977).

The Evaluation of Amalgamation Schemes

Most evaluations of structural measures emphasise the limited results that have been achieved. The principal reasons lie in the voluntary nature of most retirement, amalgamation and retraining schemes, and the usual requirement that intervention authorities work within the market mechanism for determining the ownership of farmland. More specific research problems have included the difficulty of establishing what would have happened in the absence of government intervention, and the relatively short time period for which most measures have operated.

The response of farmers to voluntary retirement schemes has been generally poor (Table 3.5) (OECD, 1972). The financial incentives of most pension and lump-sum payments have not been sufficiently large to overcome the reasons for remaining in agriculture (ADAS, 1976), and payments appear to have been made mainly to those who would have retired from farming in any event (Hine, 1973). In addition there have been marked regional variations in the adoption of retirement pensions. Thus in France the incidence of IVD payments has been greatest in central and south-western regions where the small farm problem is most acute. But within particular regions local custom (Daucé and Jegouzo, 1969), the age structure of farmers (Naylor, 1975), the impact of agricultural advisers and local lawyers (Clout, 1975), and opportunities for selling land to industrial, housing and tourist interests (Naylor, 1976) have been cited in explanation of the varied effectiveness of the payments scheme. Other problems have been posed by the restrictive terms under which financial aid is given. The original terms of the Payments to Outgoers Scheme in the United Kingdom, for example, required that the land purchased from the retiring farmer could not be resold for 40 years – a restriction subsequently reduced to 15 and then to 5 years.

The impact of farm amalgamation schemes, and the intervention authorities that operate them, has also been disappointing. The volume of land handled has been relatively small compared with the number of farms in need of enlargement and the total amount of land passing through the land market. In Sweden the area added to holdings by renting has been greater than that added by the County Boards: they have been able to supply only 30 per cent of the forest land acquired for farm enlargement (Whitby, 1968). The main problem lies in the market price of farmland which tends to exceed the maximum at which intervention agencies are permitted to purchase from the market. Their operations, therefore, tend to be limited to regions with the poorest farming and the greatest social and economic problems. SAFERs in France, for example, have been most active in southern regions where the rural exodus has been greatest and the land cheapest (Clout, 1968), whereas their activities have been limited in northern regions where land prices are higher (Naylor, 1975). Fortuitously this pattern has complemented 'remembrement' by aiding areas which have benefitted least under land consolidation schemes.

A further issue, as identified by Hirsch and Maunder (1978), is that in many countries a relatively small proportion of the land freed for amalgamation has been added to the holdings most in need of enlargement (Table 3.5). Although it may be socially desirable to assist farmers with the lowest incomes, the very nature of their position militates against them making the best use of the available land. Consequently, intervention authorities tend to discriminate in favour of middle- rather than low-income farmers (Fuller, 1975), and the land is taken by the innovators and above-average farmers in an area (Retson, 1974). Boichard (1966) recognises the economic justification for such a policy, but questions its long-term social and possibly political consequences.

The benefits of amalgamation schemes at the farm level have proved difficult to identify because data are frequently lacking on the amalgamated holding prior to the scheme. Maunder (1966) and Hine and Houston (1973) have identified increased net farm incomes following a farm amalgamation, but the increase in output per hectare tends to be less than the average achieved by farms without an amalgamation scheme. Simpson (1968) also identified a shift towards more extensive farm enterprises and a fall in the employment of farm labour following a farm amalgamation. The extent to which intervention agencies have been able to prevent farm fragmentation under an amalgamation scheme, however, has not been investigated. Similarly farmer retraining schemes have received little attention, although evidence from the Netherlands suggests that retraining has been taken up mainly by the sons of farmers with little immediate impact on farm structure.

Conclusion

Land consolidation and farm enlargement are best viewed as complementary rather than separate processes of structural change in agriculture. At one scale scattered plots of land are consolidated into more compact farms. At a larger scale farms are amalgamated to create a structure of fewer but larger holdings, albeit with an increase in farm fragmentation. Most governments promote programmes to assist these processes, but many observers are openly sceptical about the effectiveness of high-cost structural measures for solving such problems as low farm incomes and over-production (Petit and Viallon, 1970; Bienayme, 1969). Their value lies more in easing the social process of structural adjustment than in promoting it, and for success structural measures depend on the creation of off-farm employment in rural rather than urban regions. Given this perspective, structural measures need to be integrated more fully into regional rural development, with part-time farming given far more emphasis as a way of breaking the present slow exodus from agriculture (Gasson, 1977; Weerdenburg, 1973).

This analysis has identified already several aspects of structural change in agriculture that require further research. More sophisticated techniques are required, for example, to measure the spatial patterns of land and farm fragmentation, while there is scope for an updated survey of the progress made by consolidation schemes in various countries. More needs to be known about the causes of spatial variation in the spontaneous process of farm enlargement, as well as the agricultural costs and benefits of farm fragmentation. Finally, further evaluations of land consolidation and farm amalgamation schemes are required, but with attention given to both the social and economic costs and benefits of particular schemes.

Perhaps more importantly, however, attention should be turned to the issues provoked by structural changes in agriculture. A number of studies have revealed that large farms do not make a more economic use of resources than medium-sized farms (Britton and Hill, 1978; Dellaquaglia, 1978; HMSO, 1961; Lund and Hill, 1979). Efficiency is usually defined in terms of the value of outputs per unit value of inputs, while the management resource is commonly cited as limiting the efficiency of large farms (Stanton, 1978). In a British context, for example, 2-3 man farms are not substantially less efficient than 4-6 man farms, and in general there is little economic rationale for allowing very large farm units to develop. Recognising this, and placing a social value on retaining a stratum of efficient family farms, some countries, such as Denmark and Sweden (OECD, 1980), have begun to place limits on the size to which individual farms can grow. But as Smit (1975) emphasises, there are many legal ways of avoiding such limitations.

An additional issue is raised by the tendency for large farms to fall under the control of financial institutions and food processing firms (agribusinesses). Some legislatures of mid-western States of the USA have responded already by passing laws to restrict the rights of corporations to control farmland (Harris, 1980). Although much research has still to be done, there is some evidence that the managers of such farms are more motivated by purely economic goals (Heffernan, 1972), and have a lower regard for the environment (Newby *et al.*, 1977), than the family farmers whom they displace. Larger farms have the corollary of larger fields since modern machinery requires both a large total area over which to operate efficiently, and a field pattern of land in large blocks (Sturrock *et al.*, 1977). Great concern has been shown, in the United Kingdom at least, about the effects of the removal of hedgerows from the landscape in forming such a field pattern (Teather, 1970). The relationships with farm size and type of land ownership, however, have not been clearly identified.

There has also been a reawakening of interest in the social consequences of structural change in agriculture (Flinn and Buttel, 1980). With the removal of small farms from the 'farming ladder', new entrants to agriculture are increasingly the sons of farmers, farm managers, and hobby farmers who purchase land for other than farming motives. The implications of such a selective and restrictive inflow of personnel into agriculture, compared with previous circumstances, have not been fully researched. Rural sociologists in the USA, however, claim to have identified a growing two-class social system in rural areas based on large-corporate and family farm differences (Goldschmidt, 1978).

The impact of the loss of a farming base on rural communities has also emerged as an issue in recent years. Turnock (1975) has been concerned that contemporary farm enlargement is not sufficient to produce a fully viable farm system, and yet is radical enough to weaken the rural community to a point where the maintenance of essential services is no longer justified. With the decline of service provision, social and psychological pressures can be imposed on rural communities by their growing isolation.

The extent to which farm structure is stabilising at a series of optimum sizes by type of farming (Todd, 1979), or is likely to alter further, has become a matter of concern beyond agricultural production. At issue is the future pattern of land ownership, the control of farm production, the structure and stability of rural communities, and the management of the rural environment. Structural change in agriculture, therefore, is a topic worthy of further consideration by rural geographers.

References

Agricultural Adjustment Unit (1968), *Farm Size Adjustment*, Bulletin 6, University of Newcastle upon Tyne

Agricultural Adjustment Unit (1970), *Structural Change in Northern Farming*, Bulletin 14, University of Newcastle upon Tyne

Agricultural Development and Advisory Service (ADAS) (1976), *Elderly Farmers in the U.K.*, Socio-economic paper 4, Ministry of Agriculture, Fisheries and Food

Ashton, J. and Cracknell, B.E. (1961), 'Agricultural Holdings and Farm Business Structure in England and Wales', *Journal of Agricultural Economics*, 14, 472-99

Baker, A.R.H. (1961), 'Le remembrement rural en France', *Geography*, 46, 60-2

Bienayme, A. (1969), 'A propos du Rapport Vedel et du Plan Mansholt', *Review du Marché Commun*, 598-602

Binns, B.O. (1950), *The Consolidation of Fragmented Agricultural Holdings*, FAO Agriculture Studies 11, Washington

Blaikie, P.M. (1971), 'Spatial Organization in Some North Indian Villages', *Transactions of the Institute of British Geographers*, 52, 1-40

Boichard, J. (1966), 'Perspectives de l'agriculture francaise', *Revue de Geographie de Lyon*, 41, 99-127

Bowler, I.R. (1979), *Government and Agriculture: a Spatial Perspective* (Longman, London)

Britton, D.K. (1968), *The Changing Structure of British Agriculture* (Seale-Hayne Agricultural College, Newton Abbot, Devon)

Britton, D.K. (1977), 'Some Explorations in the Analysis of Long-term Changes in the Structure of Agriculture', *Journal of Agricultural Economics*, 28, 197-209

Britton, D.K. and Hill, B. (1978), *Farm Size, Tenure and Efficiency: Findings from Farm Management Survey Data for 1969-73*, Agric. Econ. Unit, Wye College, University of London

Bryant, C.R. (1975), 'The Conversion of Arable Land to Orchards: a Simulation Experiment Using Simple Probability Surfaces', *Geografiska Annaler*, 57 B, 88-99

Bultena, G.L. (1969), 'Career Mobility of Low-income Farm Operators', *Rural Society*, 34, 563-9

Bunce, M. (1973), 'Farm Consolidation and Enlargement in Ontario and its Relevance to Rural Development', *Area*, 3, 1, 13-16

Burton, S. and King, R.L. (1981), *An Introduction to the Geography of Land Fragmentation and Consolidation*, Occasional Paper 8, Department Geography, University of Leicester

Butterwick, M. and Neville-Rolfe, E. (1965), 'Structural Reform in French Agriculture: the Work of the SAFERs', *Journal of Agricultural Economics*, 16, 548-54

Calmès, R. (1981), 'L'evolution des structures d'exploitation dans les pays de la C.E.E.', *Annals de Geographie*, 90, 401-27

Carpenter, E.M. (1958), 'The Small Farmer', *Journal Agricultural Economics*, 13, 20-34

Chisholm, M. (1979), *Rural Settlement and Land Use – an Essay in Location* (3rd edition, Hutchinson, London)

Clark, G. (1979), 'Farm Amalgamation in Scotland', *Scottish Geographical Magazine*, 95, 93-107

Clout, H.D. (1968), 'Planned and Unplanned Changes in French Farm Structures', *Geography*, 53, 311-15

Clout, H.D. (1971), *Agriculture* (Macmillan, London)

Clout, H.D. (1975), 'Structural Change in French Farming: the Case of the Puy-de-Dôme', *Tijdschrift voor Economische en Sociale Geografie*, 66, 234–45

Daucé, P. and Jegouzo, G. (1969), 'The Reluctance of Farm Operators to Change their Employment', *Etudes Rurales*, 36, 37–65

Dellaquaglia, A.P. (1978), 'Size and Efficiency in Scottish Agriculture', *Scottish Agricultural Economics*, 28, 149–58

Dorner, P. (1972), *Land Reform and Economic Development* (Penguin, Harmondsworth)

Dovring, F. (1965), *Land and Labour in Europe in the Twentieth Century* (Nijhoff, The Hague)

Edwards, C.J.W. (1978), 'The Effects of Changing Farm Size upon Levels of Farm Fragmentation: a Somerset Case Study', *Journal Agricultural Economics*, 29, 143–54

European Commission (1978), *Situation et evolution structurelle et socio-economique des régions agricoles de la Communauté*, Information sur l'agric, 52 (Brussels)

European Commission (1980), *The Agricultural Situation in the Community. 1980 Report* (Brussels)

Fennell, R. (1975), 'U.K. Structural Policy and the Guidance Section of F.E.O.G.A.', *Journal Agricultural Economics*, 26, 335–49

Flinn, W.L. and Buttel, F.H. (1980), 'Sociological Aspects of Farm Size: Ideological and Social Consequences of Scale in Agriculture', *American Journal Agricultural Economics*, 62, 946–53

Found, W.C. (1971), *A Theoretical Approach to Rural Land-use Patterns* (Edward Arnold, London)

Fuller, A.M. (1975), *The Development of the A.R.D.A. Farm Enlargement and Consolidation Programme in Ontario 1966-75*, Pub. 75, Centre for Resources Development, University of Guelph

Gasson, R.M. (1969), *Occupational Immobility of Small Farmers*, Occl. Paper 13, Department of Land Economics, University of Cambridge

Gasson, R.M. (1970), 'Structural Reform and the Mobility of the Small Farmer', *Land Reform*, 2, 1–20 (FAO)

Gasson, R.M. (1977), *The Place of Part-time Farming in Rural and Regional Development*, Seminar Paper 3, Centre for European Agric. Stds., Wye College, University of London

Goldschmidt, W. (1978), 'Large-scale Farming and the Rural Social Structure', *Rural Society*, 43, 362–6

Guither, H.D. (1963), 'Factors Influencing Farm Operators' Decisions to Leave Farming', *Journal Farm Economics*, 15, 567–6

Harris, P.E. (1980), 'Land Ownership Restrictions of the Midwestern States: Influence on Farm Structure', *American Journal Agricultural Economics*, 62, 940–5

Heffernan, W.D. (1972), 'Sociological Dimensions of Agricultural Structures in the U.S.', *Rural Society*, 12, 481–99

Hill, L.D. (1962), 'Characteristics of the Farmers Leaving Agriculture in an Iowa County', *Journal Farm Economics*, 44, 419–26

Hill, R. and Smith, D.L. (1977), 'Farm Fragmentation on Western Eyre Peninsula, South Australia', *Australian Geographical Studies*, 15, 158–73

Hine, R.C. (1973), 'Structural Policies and British Agriculture', *Journal Agricultural Economics*, 24, 321–9

Hine, R.C. and Houston, A.M. (1973), *Government and Structural Change in Agriculture* (University of Nottingham)

Hirsch, G.P. and Maunder, A.H. (1978), *Farm Amalgamation in Western Europe* (Saxon House)

HMSO (1961), *Scale of Enterprise in Farming* (Natural Resources (Technical) Committee, London)

Igbozurike, M.U. (1970), 'Fragmentation in Tropical Africa: an Overrated Phenomenon', *Professional Geographer*, 22, 321-5

Igbozurike, M.U. (1974), 'Land Tenure, Social Relations and the Analysis of Spatial Discontinuity', *Area*, 6, 2, 132-5

Jackson, R. (1972), 'A Vicious Circle – the Consequences of von Thünen in Tropical Africa', *Area*, 4, 4, 258-61

Jackson, W.A.D. (1971), *Agrarian Policies and Problems in Communist and Non-Communist Countries* (University of Washington Press, Seattle)

Jacoby, E.H. (1971), *Man and Land* (Deutsch, London)

Januszewski, J. (1968), 'Index of Land Consolidation as a Criterion of the Degree of Concentration', *Geographia Polonia*, 14, 291-6

Johnson, O.E.G. (1970), 'A Note on the Economics of Fragmentation', *Nigerian Journal of Economic and Social Studies*, 12, 175-84

Johnston, W.E. (1972), 'Economics of Size and the Spatial Distribution of Land in Farming Units', *American Journal Agricultural Economics*, 54, 654-6

King, R.L. (1977), *Land Reform: a World Survey* (Bell, London)

King, R.L. (1980), 'Land Consolidation in Cyprus', *Geography*, 65, 320-4

Kingma, O.T. and Samuel, S.N. (1977), 'An Economic Perspective of Structural Adjustment in the Rural Sector', *Quarterly Review Agricultural Economics*, 30, 201-15

Kollmorgen, W.N. and Jenks, G.F. (1958), 'Sidewalk Farming in Toole County, Montana, and Traill County, North Dakota', *Annals of the Association of American Geographers*, 48, 209-31

Lambert, A.W. (1961), 'Farm Consolidation and Improvement in the Netherlands', *Economic Geography*, 37, 115-23

Lambert, A.M. (1963), 'Farm Consolidation in Western Europe', *Geography*, 48, 31-48

Lund, P.J. and Hill, P.G. (1979), 'Farm Size, Efficiency and Economies of Size', *Journal Agricultural Economics*, 30, 145-58

MacDougall, E.B. (1970), 'An Analysis of Recent Changes in the Number of Farms in the Northern Part of Central Ontario', *Canadian Geographer*, 14, 125-38

Maunder, A.H. (1966), 'Some Consequences of Farm Amalgamation', *Westminster Bank Review*, November, 57-63

Mayhew, A. (1970), 'Structural Reform and the Future of West German Agriculture', *Geographical Review*, 60, 54-68

Mayhew, A. (1971), 'Agrarian Reform in West Germany: an Assessment of the Integrated Development Project in Mooriem', *Transactions of the Institute of British Geographers*, 52, 61-76

Mayhew, A. (1973), *Rural Settlement and Farming in Germany* (Batsford, London)

Ministry of Agriculture, Fisheries and Food (MAFF) (1977), *The Changing Structure of Agriculture 1968-75* (HMSO)

Muller, P. (1964), 'Recent Developments in Land Tenure and Land Policy in Germany', *Land Economics*, 40, 267-75

Nalson, J.S. (1968), *Mobility of Farm Families* (Manchester University Press, Manchester)

Naylon, J. (1959), 'Land Consolidation in Spain', *Annals of the Association of American Geographers*, 49, 361-73

Naylor, E.L. (1975), 'The Structure of the Agricultural Population of West

Brittany', *Tijdschrift voor Economische en Sociale Geografie* 66, 159–66

Naylor, E.L. (1976), 'Les réformes sociales et structurales de l'agriculture dans le Finistère', *Etudes Rurales*, 62, 89–111

Newby, H. *et al.* (1977), 'Farmers' Attitudes to Conservation', *Countryside Recreational Review*, 2, 23–30

OECD (1965), *Geographic and Occupational Mobility of Rural Manpower*, Documentation in agriculture and food 75 (Paris)

OECD (1969), *Agricultural Development in Southern Europe* (Paris)

OECD (1972), *Structural Reform Measures in Agriculture* (Paris)

OECD (1978), *Part-time Farming in OECD Countries* (Paris)

OECD (1980), *Review of Agricultural Policies in OECD Member Countries in 1979* (Paris)

O'Flanagan, T.P. (1980), 'Agrarian Structures in North West Iberia: Responses and their Implications for Development', *Geoforum*, 11, 157–69

Parson, H.E. (1977), 'An Investigation of the Changing Rural Economy of Gatineau County, Quebec', *Canadian Geographer*, 21, 22–31

Petit, M. and Viallon, J-B. (1970), 'Reflexions sur le plan Mansholt', *Economics Rurale*, 86, 43–9

Retson, G.C. (1974), 'Farm Adjustments Associated with ARDA Farm Enlargement and Consolidation Programmes in Nova Scotia', *Canadian Farm Economics*, 9, 5, 29–38

Rickard, R.C. (1970), 'Structural Policies for Agriculture in the EEC', *Journal Agricultural Economics*, 21, 407–33

Robson, S.A. (1970), 'Agricultural Structure in England and Wales 1955–66, A Quantitative Analysis', *Farm Economics*, 11, 160–84

Simmons, A.J. (1964), 'An Index of Farm Structure, with a Nottinghamshire Example', *East Midlands Geographer*, 3, 5, 255–61

Simpson, I.G. (1968), *Change in the Structure of Yorkshire Farming*, Farm Report 177, School Agricultural Science, University of Leeds

Smit, B. (1975), 'An analysis of the Determinants of Farm Enlargement in Northland, New Zealand', *New Zealand Geographer*, 31, 160–77

Smith, E.G. (1975), 'Fragmented Farms in the U.S.', *Annals of the Association of American Geographers*, 65, 58–70

Stanton, B.F. (1978), 'Perspectives on Farm Size', *American Journal Agricultural Economics*, 60, 727–37

Steeves, A.D. (1979), 'Mobility into and out of Canadian Agriculture', *Rural Society*, 44, 566–83

Sturrock, F.G. *et al.* (1977), *Economies of Scale in Farm Mechanisation*, Occasional Paper 22, Department Land Economics, University of Cambridge

Symons, L.J. (1972), *Russian Agriculture: a Geographic Survey* (Bell, London)

Teather, E.K. (1970), 'The Hedgerow: an Analysis of a Changing Landscape Feature', *Geography*, 55, 146–55

Thompson, I.B. (1961), 'Le remembrement rurale en France: a case study from Lorraine', *Geography*, 46, 240–2

Threlfall, P. (1977), 'Government Reconstruction and Adjustment Assistance Measures in the Australian Rural Sector', *Quarterly Review Agricultural Economics*, 30, 177–200

Todd, D. (1979), 'Regional and Structural Factors in Farm Size Variation', *Environment and Planning*, A 11, 3, 257–69

Turnock, D.T. (1975), 'Small Farms in North Scotland: an Exploration in Historical Geography', *Scottish Geographical Magazine*, 91, 164–81

Vander Meer, P. (1975), 'Land Consolidation through Land Fragmentation: Case Studies from Taiwan', *Land Economics*, 51, 275–83

Vanderpol, P.R. (1956), 'Reallocation of Land in the Netherlands' in Parsons, K.H., Penn, R.J. and Raup, P.M. (eds.), *Land Tenure* (University of Wisconsin Press, Madison), pp. 548-58

Van Otten, G.A. (1980), 'Changing Spatial Characteristics of Willamette Valley Farms', *Professional Geographer*, 32, 63-71

Weerdenburg, L.J.M. (1973), 'Farmers and Occupational Change – with Special Reference to the Dutch Situation', *Rural Society*, 13, 27-38

Whitby, M.C. (1968), 'Lessons from Swedish Farm Structure Policy', *Journal Agricultural Economics*, 19, 279-99

White, J.T. (1966), 'Kerastren Bras, a private "remembrement rural" ', *Geography*, 51, 246-8

Williams, M. (1970), 'Town Planning in the Mallee Lands of South Australia and Victoria', *Australian Geographical Studies*, 8, 173-91

Williams, M. (1972), 'Stability and Simplicity in Rural Areas', *Geografisker Annaler*, 54B, 117-35

Williams, M. (1976), 'Planned and Unplanned Changes in the Margin Lands of South Australia', *Australian Geographer*, 13, 271-81

4 POPULATION AND EMPLOYMENT

A.W. Gilg

Introduction

Once upon a time the geographies of rural population and employment were one and the same thing. Nearly all rural inhabitants tilled the land or kept livestock while a small elite provided services and administration. While this pattern may still be found in the less developed world, and indeed according to Grigg (1976) about 50 per cent of the world's labour force were still engaged in agriculture in 1970, in the developed world and most notably in north-west Europe agricultural employment now occupies less than 10 per cent of the work force. In the most extreme case, the UK, only about 2 per cent of the work force are engaged in primary work, namely, agriculture and forestry. Even at such a low level it is estimated that the agricultural work force is still declining at a rate of about 3 per cent per annum (Gilg, 1981), and so British experience is almost certain to point the way ahead for other countries as they too shed their agricultural labour forces. Indeed, the Eleventh European Congress of Rural Sociology held at Helsinki in August 1981 into the theme of 'Employment: How rural is our future?' confirmed that nearly all European countries had experienced very fast rates of decline in agricultural employment since 1960, mirroring the experience of Britain in previous decades. However, loss of agricultural employment does not necessarily lead to rural depopulation, and the British model is again of interest here. During the main period of urbanisation and industrialisation, 1850–1930, there was a very rapid decline in rural population but this has been followed by a period of population growth in the countryside reversing the decline of earlier years and providing a dichotomy between population and employment, so that in many areas the 20 per cent of the population who live in the countryside do not necessarily work there. Because of this functional split, the chapter is divided into two main sections, dealing with population first and then with employment.

74

Rural Population

Because populations are so complex and dynamic, both through space and time, their study presents special problems, not least in the collection of basic data. Accordingly, much work is needed even to provide a simple description of population and its structure. This means that little time and resources have been left for explanation. It is therefore appropriate to begin by examining general descriptive studies at three spatial scales (the world, the country and local levels) before looking at some of the explanatory models that have been developed in the last decade.

Descriptive Work in Rural Population Geography

The World Scale

At the outset it must be said that there are few if any studies at the world scale that look at rural populations specifically. General texts on population geography tend to devote only a few pages to rural population. For example, Zelinsky (1970) confines himself to half a dozen pages which mainly show how rural populations are inferior to and dependent on an urban dominated civilisation. Another and more recent book on population geography (Jones, 1981) uses a process-oriented and socially relevant approach. Neither book sees the rural-urban split as a major explanatory variable, and perhaps this is not surprising since it has become increasingly difficult to differentiate between urban and rural populations in their social and economic characteristics. Some workers prefer to see a rural-urban continuum rather than a decisive divide (Lewis, 1979) while others reject the difference as irrelevant. This approach has undoubtedly been influenced by text-books on demography, for example, Pressat's *Statistical Demography* (1978) which is concerned almost entirely with the dynamics of population.

This process approach is reflected in perhaps the best known and respected text, Beaujeu-Garnier's *Geography of Population* (1978), which devotes nearly half of its 381 pages to natural population change and migration but only 15 pages to rural population, and most of these comprise only passing comments. For example, the notion of rural and urban population is described as unsatisfactory since different countries use different administrative definitions for rural populations. Accordingly, Beaujeu-Garnier considers that the only basically rural societies are those where 'the population lives essentially on what the local soil produces', and although the areas of the world with primitive societies are shrinking, Beaujeu-Garnier defines a list which still includes nearly all of Africa, central America, northern Brasil and parts of southern Asia and Indo-China. But rural populations can be set aside from urban populations by another characteristic, namely, fecundity. Beaujeu-Garnier produces figures to show that birth rates are higher in the countryside in both the developed and less developed world. Population growth by migration is, however, not common to both regimes. In the developed world the last two decades have seen a net 'de-concentration' of population as people have moved out not only into new suburbs but into country towns and villages as well. In the less developed world though, cities are growing at very rapid rates and the world's largest and fast growing cities are now found in the Third World (Pacione, 1981). Grigg (1980), however, points out that although much attention has been paid to rural-urban migration in the Third World, in some countries rural-rural migrants have been the more numerous.

At the world scale, therefore, two broad types of rural population can be discerned, the 'primitive' and the 'ex-urbanite', with, of course, major bands of transition between the two extremes.

The Country Scale

Because Britain has one of the oldest census records and the longest history of rural depopulation, followed by rural repopulation, this section uses Britain as a case study. Since the first decennial census was held in 1801, the only exceptions being 1941 (not held because of the war) and 1966 (a 10 per cent sample census), much experience in handling census data and presenting it in an easily assimiliable form has been obtained (Saville, 1957; Smith, 1951; Vince, 1952; Willatts and Newson, 1953). Clearly not all this work has been done by geographers and much of it has looked equally closely at urban as well as rural areas. Indeed, much of the work has been done by the staff of the Office of Population Censuses and Survey (OPCS), formerly the Registrar's General Office. Though much of their data are not put into their spatial context, the government often only being interested in broad trends, a great deal of work (Hakim, 1978) has been done on small area statistics, namely parish data, and ways in which they can be analysed.

The map remains a basic form of description, particularly in showing different rates of change and for depicting variations in population change at the sub-regional scale, for example, by age group. For instance,

Champion (1976) in an analysis of population change between 1951 and 1971, uses maps of natural change and net migration to show that: 'the most dramatic differences were found in the South West. Devon and Cornwall emerged from a period of relative stagnation into one involving population growth rates at almost twice the national average. Reference to the breakdown of total change into net migration and natural change indicates the role of the accelerated influx of more elderly people in producing this reversal.' Elsewhere, Champion (1973) uses tables to describe the decentralisation of population and the repopulation of rural districts and small towns as shown in Tables 4.1 and 4.2.

Although these two tables show the most rapid rate of population growth to be in England's two most rural regions, the South West and East Anglia, and that population growth was most rapid in rural districts

Table 4.1: Regional Variations in Population Growth Rates, 1951-71

Region	% Growth	
	1951-61	1961-71
Northern	3.6	1.4
Yorkshire and Humberside	2.5	3.5
North-West	1.9	2.7
East Midlands	7.1	9.4
West Midlands	7.6	7.4
East Anglia	6.4	13.6
South-East	7.6	5.9
South-West	5.6	10.9
Wales	1.8	3.3
England and Wales	5.4	5.7

Source: Champion, 1973.

Table 4.2: Population Change, 1961-71, by Settlement Category

Settlement category	Population change	Rate %
Conurbation	−813,910	−4.9
Areas outside conurbations		
Urban areas with populations:		
over 100,000	+ 113,657	+ 1.7
50,000-100,000	+ 373,509	+ 7.5
under 50,000	+ 1,201,478	+ 13.7
Rural district	+ 1,614,376	+ 18.0
England and Wales	+ 2,489,110	+ 5.4

Source: Champion, 1973.

and small towns with a population below 50,000, they do tend to over-emphasise the position. Firstly, the South West and East Anglia are not homogeneous regions and most of the growth has taken place in a ring around London, or put another way in an arc stretching from Ipswich through Peterborough to Swindon, Bristol and Bournemouth. Northern parts of East Anglia continued to lose population. In other words, the growth is largely due to people seeking a pleasant life-style in the villages and small towns of central southern England and the outer Home Counties. Another way in which the figures can distort is if too coarse a spatial unit is employed. For example, much of the population growth registered for rural districts in Table 4.2 is in reality growth adjacent to a much larger urban area and therefore not truly rural.

Another problem with census data lies with their decennial collection and as Champion has shown a lot can change in ten years. Annual population estimates are collated by the Office of Population Censuses and Surveys using the register of births and deaths and the electoral register but these are only published for post-1974 district council areas and so show only broad trends. Another estimate can be obtained from the records of Area Health Authorities and Kennett and Spence (1979) used these to show that the trends of the 1950s and 1960s outlined by Champion were not only continuing but increasing in intensity and spatial impact. For example, parts of rural Wales began to show population growth for the first time for over a century, and Powys became the eighth fastest growing county with a growth rate of 6.78 per cent.

The use of the map in rural population work received a great boost with the advent of three innovations in the 1970s; first the widespread availability of computers with the massive stores needed to handle the vast amounts of data generated by a population census; second, the decision by the OPCS to make the 1971 and 1981 censuses available in computer readable forms, normally magnetic tape; and third, the decision to record census data by national grid reference and to publish data for different sizes of grid square, the smallest for rural areas being 1 kilometre squares. At a stroke these developments removed two of the most difficult constraints, the time taken to map population characteristics and the coarse and arbitrary spatial pattern of administrative units. Now admittedly after a good deal of lengthy data preparation, thousands of different map permutations at a very fine spatial scale can be produced in the space of a working week.

There have been problems however. The OPCS was very slow in producing machine readable data and the sheer amount of the data, the poor definition of line printers and the fact that mapping programmes like SYMAP are rather slow in their execution meant that the 1971 census was not fully computerised till the late 1970s (Kirby and Tarn, 1976). Gradually, however, new techniques and technologies were developed, for example, laser beams to replace the line printer (Rhind, 1976). Much of this work was by definition long term, and so the OPCS, the Institute of British Geographers and the Social Science Research Council encouraged one particular department of geography, at Durham, to set up a major project into mapping the 1971 census by computer under the organisational title of the Census Research Unit (Rhind, 1975).

In 1980 the Unit completed what can only be described as a major innovative breakthrough when they published in association with OPCS and the General Register Office (Scotland), 'People in Britain: A Census Atlas' (Census Research Unit, 1980). Although the data used in the Atlas were already nine years out of date when published, the technological progress made at the very edge of computer capacity should mean that the 1981 and subsequent censuses will be mapped in much quicker times.

The Atlas is divided into two parts, 34 maps covering all of Great Britain and seven sets of regional maps covering the main conurbations. Each map is accompanied by a short commentary. For rural geographers, the most useful map is that for population density based on 1 kilometre squares, which shows for the first time the fine-grained pattern of rural population density. Other useful maps are those showing farmers and journeys to work. However, the Atlas's most useful contribution is that it shows what can now be achieved and points to the enormous potential work that can be done on the 1981 census.

At the time of writing, however, six months after the April 1981 census only the basic population totals for each district have been published (Office of Population Censuses and Surveys, 1981; Registrar General, Scotland, 1981). None the less, some useful statements can be made about recent population changes even from these preliminary data. First of all the population has virtually stopped growing with an increase of only 0.5 per cent over the decade. Second, the regional variations in growth or decline first noted in the 1960s and due to migration have continued into the 1970s, as shown in Table 4.3. The South West and East Anglia have continued to be the fastest growing areas (outside the rural South East), the South West absolutely and East Anglia relatively. But, as pointed out earlier, this is mainly due to a flight from overcrowded and congested cities and to restrictive

Table 4.3: Regional Population Change, 1971-81

Region	Change (000)	%
North	−45	−1.4
Yorkshire and Humberside	−2	−0.1
East Midlands	+ 174	4.8
East Anglia	+ 196	11.7
South East (London)	−756	−10.1
South East (outside London)	+ 555	+ 5.9
South West	+ 246	+ 6.0
West Midlands	+ 27	0.5
North West	−191	−2.9
Wales	+ 49	+ 2.2
England and Wales	+ 262	+ 0.5

Source: Office of Population Censuses and Surveys, 1981.

planning policies preventing the spread of conurbations (Hall *et al.*, 1973).

Interestingly, the three biggest increases of between 60 and 85 per cent in Milton Keynes, Redditch and Tamworth were all associated with New Towns being built in the open countryside. In line with this trend another noteworthy feature of the 1970s is that jobs have now started to decentralise in significant numbers as well, but it will be the mid-1980s before census data will be able to put an exact figure on this.

Taking the analysis down from the regional to the county level also shows some new trends as shown in Figure 4.1. Two types of rural growth area are found − first, counties within 50-75 miles of a major conurbation, for example, Shropshire and Sussex, and second, counties that are not only more rural but far more remote, for example Cornwall, Norfolk and Powys. These latter changes are difficult to explain and will need further data before more than a guess can be made. Others are more easily explicable, for example, commuting overspill in Somerset, and New Town expansion in Buckinghamshire and Shropshire.

Analysis of population change at the district level provides some further clues. The OPCS has classified districts into eleven types and, as Table 4.4 shows, apart from New Town districts, rural districts have shown the largest relative growth, and overall, the largest absolute growth. The most interesting feature of this table is, however, that about 30 per cent of the British population now appears to live in semi-rural or rural areas, compared to the commonly accepted figure of 20 per cent.

Figure 4.1: Percentage Population Change, 1971-81, England and Wales

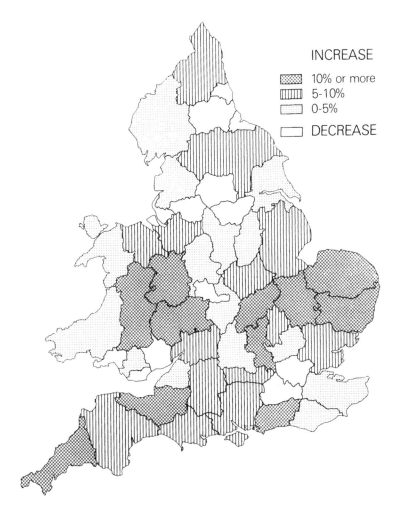

    Similar results are not however reported from Scotland (Registrar General, 1981). Here the percentage of the population living in the Central Clydeside Conurbation has risen slightly from 33.0 per cent in 1971 to 33.5 per cent in 1981, while the greater rural region, Strathclyde, which also includes Glasgow, has fallen from 49.3 per cent to 46.9 per cent. The southern regions, Borders, and Dumfries and Galloway have barely increased their share and though the two northern

Table 4.4: Population Change by Type of District, 1971–81

Category of district	1981 (millions)	Number	Change (000)	%	Decrease over 5%	Decrease 0-5%	0-5%	Increase 5-10%	over 10%
England and Wales	49.0	403	262	0.5	55	71	92	88	97
Inner London	2.5	14	−535	−17.7	13	–	–	–	1
Outer London	4.2	19	−221	−5.0	9	10	–	–	–
Major cities	3.5	6	−386	−10.0	5	1	–	–	–
Other cities	7.7	30	−160	−2.0	7	11	12	–	–
Other cities over 175,000	2.8	11	−149	−5.1	4	6	1	–	–
Smaller cities	1.7	16	−55	−3.2	6	5	5	–	–
Industrial districts	6.7	73	200	3.3	6	18	24	16	9
New Towns	2.2	21	283	15.1	1	1	4	3	12
Resort and seaside retirement	3.3	36	156	4.9	2	3	11	12	8
Other urban, mixed urban-rural and more accessible rural districts	9.4	99	661	7.8	2	14	21	28	34
Remoter largely rural districts	5.0	78	468	10.3	–	2	14	29	33

Source: Office of Population Censuses and Surveys, 1981.

regions, Grampian and Highland, have gained population this is almost certainly due to the short-term impact of North Sea oil.

It has already been emphasised that the sheer quantity of census data leaves little time for analysis, but some work has none the less been done at a national scale. For example, Cloke (1978), using nine variables from 1961 and 1971 census data, constructed an index showing the degree of rurality for local authority areas in 1961 and 1971. By comparing the indices Cloke produced a typology of rural areas that were: becoming increasingly non-rural, were static, or becoming increasingly rural. Cloke then proposed a cyclic process for rural change with reversed depopulation and reversed growth as two poles of the cycle. This follows the same lines as Woodruffe (1976), who divided rural districts into six types depending on their history of population change:

1. Accelerated depopulation.
2. Reduced depopulation.
3. Reversed depopulation.
4. Reversed growth.
5. Reduced growth.
6. Accelerated growth.

In spite of the development of theories and the recent possibility of testing them at national level using computer techniques, many geographers prefer to work at the local level and come to grips with real people via questionnaire and other work rather than relying on the massive and infrequent data provided by the census. Accordingly, the next section considers work done at the local level.

The Local Scale

One of the advantages of work at the local scale is that data can be collected to test specific hypotheses. For example, Bracey (1958) in a study of 375 Somerset parishes related rural depopulation to the provision of services and found:

a) parishes which have experienced persistent depopulation over a long period tend to occur in areas which are relatively remote from well-populated urban areas;
b) the number of services per parish in these areas is, in general, lower than in areas where population is increasing;
c) for each group of services there are parishes in areas of increasing

population which have as few services or organisations as parishes
which are declining in population;

d)   parishes in depopulating areas are, on the whole, much smaller than
those in areas of increasing population, and this fact seriously
influences the development and maintenance of new services and
organisations of all kinds.

Johnston (1967) in another area study tested Ravenstein's first law
of migration which states that the great majority of moves take place
over short distances. Because the census did not provide data for short-
term moves until the 1961 census onwards, this theory had remained
untested since its formulation in 1885. Instead, Johnston used electoral
register data for 21 electoral districts to see how many people had
moved between 1951 and 1961, the proportion of movers to stayers,
and the direction of short-term migration. He found that though the
net change was small, the gross change had been large, and that by using
a migration quotient:

$$\text{Migration quotient} = \frac{\text{Total residents at any time 1951-61}}{\text{Average of 1951 and 1961 population}}$$

he concluded that a migration quotient of 2.00 indicated a complete
turnover of population within the ten-year period. No less than 11 out
of the 21 districts recorded a quotient of 1.80 or more, showing the
value of such detailed work. However, Ravenstein's first law was not
fully vindicated since only one in five of all moves were local, but this
may have been due to shortcomings in the data base or the special
circumstances of the area. Indeed, the tedious and slow nature of the
work prevented Johnston from improving the data base by taking into
account deaths and marriages, and these difficulties are the most potent
reasons why large studies of local rural populations using their own data
are somewhat infrequent.

It is not surprising therefore that one of the classic studies of rural
population remains Jackson's (1968) census study of the North Cotswolds. This study took advantage of two new features in the 1961
census, the increasing coverage of the questions asked (using 10 per
cent sampling procedures) and the availability of detailed data at the
parish level. This allowed Jackson to produce a detailed analysis of
population, housing, migration and employment for the area. Her main
conclusion was that the majority of the area had lost population and
would continue to do so unless an employment alternative to agriculture could be found. Masking the loss was the substitution of retired

people moving into the area and replacing the younger people who had left in search of jobs. So interesting and detailed was the work that the area was chosen for a major restudy by the University of Birmingham and Wye College in the late 1970s. (This work is reported elsewhere in Alan Roger's chapter on housing.)

An alternative to an area study is to look at villages or even one village. But here again the amount of time involved prevents many geographers from even contemplating the task. Not surprisingly, therefore, most of the main studies have been conducted by county councils for their structure plans or for other purposes. One of the best examples is a 1965 study of Hampshire villages involving 1,713 interviews (Hampshire County Council, 1966). Such local studies allow motivation to be studied and the Hampshire study found that there were significant differences given for moving into or out of a village depending on age and social class, with a tendency for young working-class people to leave the villages and for older professional people to move into villages. Further light on the social aspects of local population change has been provided by Pacione (1980) in his study of Milton of Campsie, near Glasgow. In this rapidly expanding village (75 per cent in seven years) two main types of migration had occurred, first into 1950s council housing, and second into 1970s private housing estates. Pacione's main findings were that 85 per cent of the council tenants had been born locally but only 18 per cent of the private owners, and that only 10 per cent of council tenants worked more than two miles away from the village, but that 66 per cent of private owners did. This indicates an enormous difference in mobility between social classes, and emphasises once again the increasing complexity of population characteristics when an ever finer spatial scale is used.

Summary of Other Descriptive Work

Although most of the studies discussed above have been conducted in Great Britain, much rural population work has also been conducted elsewhere, for example Ogden (1980) and White (1980) have examined rural depopulation and the changing structure of rural communities in France, and Buksmann (1980) has studied agricultural migration in Bolivia. A recent conference (Lonsdale and Holmes, 1981) on sparsely settled regions in Australia and the USA produced six generalisations thought to be generally applicable to most rural regions. These were that 1) the level of service provision is crucial, and that 2) since governments are committed to regional equity that they 3) subsidise essential services but 4) only up to a certain point and 5) that one of the conditions of living in a sparsely populated area is a lack of service provision. The sixth generalisation is different and states that sparsely-populated regions are more dependent on external forces, mainly the price paid for their primary products. These generalisations have begun to introduce an explanatory element into the discussion and so the next section looks in more detail at explanation in rural population geography and other developing themes of the last decade.

Explanatory Work and Emerging Themes in Rural Population Geography

Density and Population Potential

It has already been pointed out that the map has been the traditional form of description in rural population geography, and it is only one short step from the map to the concept of population density. Density figures can be very revealing; for example they show that: 'the greater part of the physical area of the country is inhabited at relatively low densities' (Office of Population Censuses and Surveys, 1980). Put another way, 25 per cent of the population are found spread over 90 per cent of the land area, or that while the average density is 2.4 people per hectare the median density is only 0.5 people per hectare. In other words Britain is areally speaking still a very rural country, and from Table 4.5 it appears to be becoming increasingly so, with more people living in very sparsely populated or virtually uninhabited areas. But closer analysis (Craig, 1979) reveals that it is the areas of suburban density that have experienced the greatest overall growth in both population and extent.

Best and others have further developed this theme by using regression analysis to propose a density-size rule which states: 'As the population size of settlement increases, the land provision declines exponentially' (Best *et al.*, 1974). For the whole country they found that densities were however tending towards a pivoted density with metropolitan areas shedding their populations and small country towns increasing their densities by infilling. In a totally rural study using land-use maps from the Second Land Utilisation Survey and the development plans of local planning authorities, Best and Rogers (1973) produced a least squares regression model of the form:

$$\log y = 1.9086 - 0.1522 \log x$$

Table 4.5: Changing Densities by Ward/Parish, 1931–71, Great Britain

Density persons/hectare		Population (000)				Area (000 hectares)			
		1931	1951	1961	1971	1931	1951	1961	1971
0–0.02	Virtually uninhabited	18	21	28	30	1,491	1,772	2,132	2,450
0.02–0.10	Very sparse	219	230	249	253	4,498	4,523	4,664	4,780
0.10–0.20	Sparse rural	502	450	501	495	3,472	3,121	3,338	3,342
0.20–1.50	Dense rural	5,141	5,171	4,887	4,654	10,572	10,169	9,305	8,490
1.50–25.00	Industrial rural and suburban	14,570	17,116	18,618	21,408	2,427	2,784	2,858	3,181
25.00–100.00	Urban	13,343	18,946	21,975	24,279	306	414	498	565
100.00 plus	Dense urban	11,002	6,920	5,026	2,859	63	47	35	22
	Total	44,795	48,854	51,284	53,978	22,829	22,830	22,830	22,830

Source: Office of Population Censuses and Surveys, 1980.

where log y =   total land provision (ha/1000 people)
and   x =   population of settlement

for settlements with a population between 10 and 10,000 people. By plotting deviations from the regression, Best and Rogers were able to show those areas with low densities and thus potential for infilling and those areas with high densities where areal expansion should be allowed.

In spite of this useful work, density remains a rather static and misleading statistic in that it depends very much on the unit used to compute the density. An alternative but related statistics, population potential (Craig, 1972), has therefore been developed to take into account the catchment area of population. None the less, if different spatial units are used considerable variations in resulting potential distributions can be found (Craig, 1974). However in remote rural areas the related concept of accessibility (Robertson, 1976) can be really useful in planning the optimum location for services and thus help to retain the area's population, especially if census grid squares and real ground distances are used.

Population density in the form of settlement size has also been frequently used to explain the level of services that an area can support in the form of a hierarchical threshold. For example, Edwards (1971) in a study of north-east England found the following hierarchy:

Medium market settlements   411 to 900 adults
Agricultural/service settlements 165 to 410 adults
Small villages   90 to 164 adults
Hamlets   36 to 89 adults
Small hamlets and farmsteads   Fewer than 36 adults.

Other studies have confirmed the existence of rural central place hierarchies along the lines of classic central place theory. But with increasing mobility most of these settlements are now redundant and with depopulation further reducing their population potential their future is very problematical. Indeed the troubled future of rural settlements following rural depopulation is one of the main reasons why the next approach to rural population geography, the study of rural migration, has been dominated by rural depopulation while rural repopulation in rural lowland England has been almost completely neglected (Edwards, 1973).

Rural Migration

It is very interesting to note that the recent repopulation of the countryside is the mirror opposite of Ravenstein's (1885) second law of migration:

> The inhabitants of the country immediately surrounding a town of rapid growth flock into it; the gaps thus left in the rural population are filled up by migrants from more remote districts, until the attractive force of one of our rapidly growing cities makes its influence felt, step by step, to the most remote corner of the kingdom.

Ravenstein used the censuses of 1871 and 1881 to formulate his seven laws of migration, and the census has remained a main source of data for measuring rural depopulation. But with greater mobility the decennial census became too infrequent a source, and so in the 1960s and 1970s, a number of new sources were developed (Edwards, 1973), including the use of electoral registration lists, the National Health Service central register and Department of Employment National Insurance Card records. It has already been noted that Johnston used electoral register data in his mid-1960s study of Nidderdale. Similarly, Jones (1965) used electoral registers to test Ravenstein's laws in central Wales and found them to be broadly true. Ten years later Jones (1976) used a different technique in a study of migration to a planned growth point, Newtown, which had been selected to reverse long-term population decline in mid-Wales. Jones interviewed 184 in-migrant households and found that the following factors helped to explain the pattern of in-migration.

1. A Welsh background, youthfulness and previous migration experience, contribute to migration potential.
2. A knowledge of destination gained from previous visits or from friends and relatives, fashions search procedures.
3. Employment and environment, count for more than housing in an evaluation of place utilities.

Newtown has been used as a growth pole to maintain the population of mid-Wales, while accepting that the countryside around it will continue to depopulate. Similar conclusions have been reported in north-east Scotland (Turnock, 1968) and another study of very rural areas (Treasury, 1976) concluded: 'Some greater degree of concentration is probably essential if depopulation is to be checked', after pointing out the greater costs for private individuals and public services of living and operating in remote, sparsely settled areas.

It should already be clear that migration is a very complex process with many interacting variables. Hannan (1969) found differentials due to sex, education and distance from large towns, and Johnston (1971) found that those with extensive kinship networks were less likely to migrate. Not surprisingly most models of rural population change have been rather simplistic, for example, Johnston (1965) concluded that there had been two major factors which had had a considerable influence upon the type of population change in English rural areas, suburbanisation and the settlement pattern. However, there have been attempts to provide a general model. For example, Mabogunje (1970), using African experience where rural-urban migration is now reminiscent of earlier migrations in Europe, proposed a theory of rural-urban migration based on a General Systems Theory framework. Mabogunje argued that one of the major attractions of this type of approach was that it sees rural-urban migration 'no longer as a linear, uni-directional, push-and-pull, cause-effect movement but as a circular, interdependent, progressively complex, and self-modifying system in which the effect of changes in one part can be traced through the whole of the system. In spite of Mabogunje's arguments for a general model, most work has continued to concentrate on rural depopulation and a recent book on migration (White and Woods, 1980) almost totally ignores the selective repopulation of the countryside that has been such a characteristic of the last 15 to 20 years in the developed world. However, interest in migration seems to have waned a little in recent years and increasing emphasis has been placed on classifying the less dynamic aspects of population.

Classification as an Explanatory Tool

Three trends culminated in the 1970s to give a great impetus to the classification of rural populations into different typologies or groupings; first, the advent of really powerful computers, second, the publication of census data on computer tape and third, the widespread availability of package programmes for multi-variate statistical techniques like factor, cluster and principal components analysis. None the less, some of the models were relatively simple, for example, Craig and Frosztega (1976) tested two assumptions: that density declines with distance from the centroid of a population centre and that the populations of these centres follow some regular pattern

Table 4.6: Rural Population as Classified by OPCS

Family and cluster	Number of districts	Per cent of GB population
F1 Suburban and growth areas		
C2 Rural growth areas	31	4.8
F2 Rural and resort areas		
C7 Rural Wales and Scottish Islands	16	1.1
C8 Rural West	32	3.5
C9 Rural East	31	4.0
C10 Rural Scotland	23	1.6

Source: Office of Population Censuses and Surveys, 1978.

(eg, rank size or log-normal). They found an increasing fit from 1931 to 1961.

Much more complicated work has been conducted, however, by a grouping of workers from the OPCS, Planning Research Applications Group and the Centre for Environmental Studies (Office of Population Censuses and Surveys, 1978). Using 40 variables from the 1971 Small Area Statistics data they classified the British population into six spatial family types with a number of sub-clusters, making 30 groups altogether. Five main rural areas were found, as shown in Table 4.6 and Figure 4.2.

The rural and resort areas were characterised most of all by a high concentration of self-employed and agricultural workers, and small proportions in the 15–24 age group. Socio-economically the rural areas are close to the national mean but are over-represented by professional/managerial and semi-skilled workers. This dichotomy between rich and poor in rural populations is brought out by other variables, for example, an above average share of owner-occupied housing being counterbalanced by an above average share of unfurnished privately rented accommodation (tied cottages).

Although these classifications and their areal spread tend to confirm the generally accepted view of rural population geography, the authors admit that multivariate classification is a subject on which virtually no two practitioners agree, but argue that the crucial test should not be about the methodology used but whether the classifications are useful.

None the less, in a critique of the OPCS classification, Openshaw, Cullingford and Gillard (1980) point out that statistical classification

Figure 4.2: OPCS Classification of Rural Areas

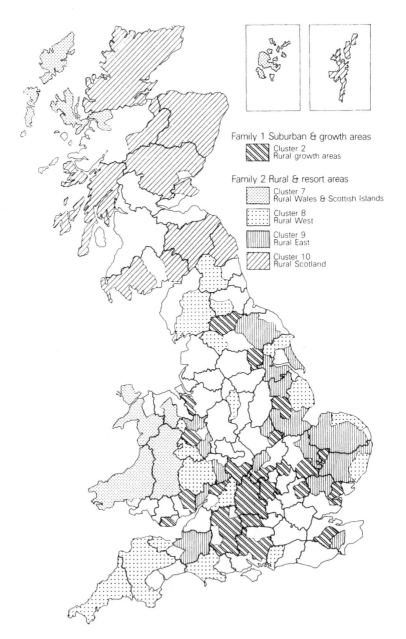

Family 1 Suburban & growth areas
Cluster 2
Rural growth areas

Family 2 Rural & resort areas
Cluster 7
Rural Wales & Scottish Islands
Cluster 8
Rural West
Cluster 9
Rural East
Cluster 10
Rural Scotland

methods are by their very nature exploratory data analysis techniques, heavily dependent on the methods used and on a number of arbitrary decisions which should reflect the purpose of the classification. But the OPCS classification, claims Openshaw, had no single purpose in mind. Another problem is raised by the large numbers of zones that have been classified. Openshaw also points out that few taxonomic exercises of this size have been attempted and it is therefore inevitable that the results should be seen as a pioneering first attempt, rather than as a polished final product based on the application of well-established proven methods. In replying to Openshaw's criticisms the OPCS reiterated the difficulty of working with such vast data sets and with new complex statistical procedures involving dozens of variables, and claimed that its classification is as far advanced as any could be bearing in mind the constraints of public policy, and the realities of time and cost (Webber, 1980). Clearly these techniques are still in dispute and there may still be a doubt as to whether they are really cost-effective in the degree of explanation they afford.

Not all population models deal with classification problems however. Rees and Wilson (1976) try to integrate demography with spatial analysis, and in a highly mathematical text they concentrate on the way in which population can be divided into demographic states and then the way in which there may be transitions from one state to another. This work is conducted at the multi-regional level and has no specific section on rural population. Furthermore, in a test of four models of population change, Rees and Wilson (1975) are looking for a 'truly general model of population growth' and again ignore the rural-urban variable. It is of course possible that rural population geography is synonymous with urban population geography, but this ignores many of the differences already described in this chapter.

Rural Employment

The Factual Background

It is very difficult to obtain accurate figures for truly rural employment for two reasons. First, the areal units used often contain substantial towns within their borders, and second, the annual and monthly data collected by the Department of the Employment (DoE) only record 'employees in employment'. The following categories of workers are excluded: working proprietors, partners, the self-employed, wives working for husbands, husbands working for wives, persons working in their own homes, former employees on the payroll as pensioners only, and private domestic staff working in private households. In rural areas these categories are often over-represented — for example, most farmers are self-employed, their wives work for them or as partners, professional people may often work from home preferring a rural to an urban base and large estates still employ a number of private domestic staff. In an extreme case like the Western Isles of Scotland the official employment figures for June 1977 showed that 6,700 people were in employment and 1,000 unemployed, giving a total of 7,800. But this employment figure represented only 26.4 per cent of the population (McCallum and Adams, 1981), compared to a comparable figure of 42 per cent for Scotland as a whole. By sector the discrepancies are just as vast, for example, the Department of Agriculture for Scotland (DAFS) estimated there to be 75,000 people working in agriculture in Scotland in 1976, but the DoE data only revealed 33,000 people or just 44 per cent of those estimated by DAFS.

What is certain, however, is that employment in the primary sector has fallen for over a hundred years and continues to fall by around 2 to 3 per cent a year. This is not necessarily due to poverty but instead to increasing agricultural efficiency. For example, Drudy (1978), in a study of a prosperous farming area, north Norfolk, found that changes in agriculture were the foremost reason why the population of the area had decreased by 3.7 per cent between 1951 and 1961 and by 6.9 per cent between 1961 and 1971. A study of 153 farms found that redundancy was the most important single reason for leaving farm employment between 1960 and 1970, as shown in Table 4.7.

Drudy also found evidence that depopulation had been worse in small settlements and concluded that a low level of social provision added to the lack of a sufficient employment alternative to agriculture could lead to a vicious circle of decline based on the theory of cumulative causation, as shown diagrammatically in Figure 4.3. Hodge and Whitby (1981) concur with the employment aspect of this theory and advance as their main thesis the view that rural labour markets are at the core of the problem of rural depopulation in developed countries.

Similar findings have been found in a marginal region, the Galway area of Ireland, not only in relation to workers leaving agriculture, but to young people leaving the area before they even enter the job market. Many of the young people who leave are the most able and motivated, adding weight to the theory of cumulative causation, as the less able young, the old and the redundant come to form a greater and greater

Table 4.7: Reasons for Mobility from Norfolk Farms, 1960-70

	Number	%
Redundancy	154	32.3
Low wages	111	23.3
Dissatisfied with work or conditions	86	18.0
Dismissed	35	7.3
Attracted by other employment	24	3.0
Other reasons	67	16.1
Total	477	100.0

Source: Drudy, 1978.

Figure 4.3: The Cyclic Nature of Rural Decline

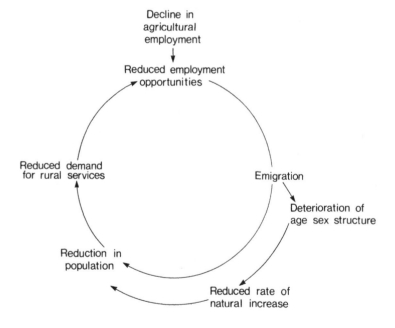

proportion of the population. Accordingly, Drudy and Drudy (1979) advocate the provision of professional and office jobs as well as manufacturing jobs as an alternative to agricultural employment, so that the aspirations of the whole ability range of young people may be met (Drudy, 1975).

One section of the farming community, the farmers themselves, has however been more resistant to migration, and because farm workers have left, until recently, at a faster rate than farmers, there are now about 210,000 full-time farmers in the UK compared to about 180,000 full-time workers. Farmers are very reluctant to leave farming as a number of studies have shown (Gasson, 1969) and their numbers are only now beginning to fall as an increasing percentage reach retiring age, but as they do so it is feared that in some areas it will be very difficult to find replacements.

In summary, agricultural workers and now farmers too, have been a rapidly declining feature of the labour force in rural areas. In the remote and marginal areas this has led to actual population loss, but in lowland areas accessible to major towns the suburbanisation of the countryside has masked the decline by providing numerous service jobs, and increasingly, decentralised work as employers seek pleasanter and more accessible work locations in the small towns and large villages of lowland England. The real employment problem is thus to be found in the remoter and upland areas and most research work in the last decade has centred not upon the facts of employment decline, but on possible policies to reverse the trend (Gilg, 1976) and the unbalanced nature of employment in these areas.

Rural Employment Policies and Proposals

There have been almost as many proposals for solving the problems of rural employment as there are people involved. But they do have a number of common themes. For example, the Association of District Councils (1978) believe the following needs ought to be satisfied:

   a) relief of unemployment and creation of employment opportunities for small-scale factory developments and promotion of tourism;
   b) provision of new housing and rehabilitation of existing rural housing;
   c) improvement of public transport;
   d) development of infra-structure including water and sewerage schemes;
   e) provision of adequate local community facilities.

The Country Landowners Association (1980) concluded that:

   c) There is a requirement for alternatives to agricultural employment if a healthy rural community is to be maintained along with the necessary public services;

d) encouragement of small industries benefits the country as a whole as such industries create the majority of new employment opportunities . . .

h) Local authorities should be encouraged to allow the conversion of redundant farm buildings to small work shops and holiday accommodation;

i) the work of COSIRA should be expanded.

The National Council for Voluntary Organisations (1981) called for a simplification of bureaucratic procedures, closer liaison and partnership between government agencies and those in the private and commercial sector, flexibility in planning controls, greater job creation by larger landowners, rural estates, and major private and commercial interests in rural areas, and more communal self-help since rural areas represent widespread opportunities for initiative in job creation by cooperatives, local enterprise trusts, community groups and others.

In a more analytic approach an inter-departmental group of civil servants from nine government departments looked at six rural areas and estimated that for a rural population of 2½ million an extra 2,500 jobs would be needed each year (Treasury, 1976). They concluded that the best hope of preventing depopulation lay in policies designed to encourage the growth of light industry mostly in small towns and that the main requirements for this approach were:

a) more emphasis on the provision of housing, advanced factories and urban services generally, and

b) organisations for providing assistance which are in touch with local conditions and able to concern themselves with quite small projects.

These conclusions were based on the public expenditure costs per job shown in Table 4.8.

Although the study admitted that the costings could easily be altered by changing circumstances or assumptions, it concluded that only manufacturing industry was likely to provide resource gains. Hill farming could produce gains but was unlikely to do so because of increasing mechanisation and forestry was seen as a total non-starter.

On a wider dimension the Council of Europe (1976) has advocated: the attraction of manufacturing industry; improvements to the basic infrastructure; the development of small-scale and local enterprises based on local resources; the development of recreation and tourism;

Table 4.8: Different Costs for Providing Various Rural Employment Options

	Public expenditure costs				Net resource benefits/efforts		
	Light industry	Hill farming	State forestry	Private forestry	Hill farming	Forestry	Light industry
Rural Northumberland	1,300	4,000	29,000	12,500	1,000	−27,000	1,270
Merioneth	1,280	3,700	29,000	8,000	1,100	−23,000	810
Montgomery	1,285	2,200	29,000	8,000	300	−23,000	810
Rural Camarthen	1,305	4,300	23,000	8,000	1,700	−23,000	820
Sutherland	1,300	3,400	30,000	13,000	300	−23,000	770
Roxburgh	1,350	1,300	32,000	12,500	500	−22,000	300
Wigtown	1,250	7,500	29,000	11,000	2,600	−19,000	1,650
Average	1,300	3,800	28,700	10,700	1,050	−22,800	920

Source: Treasury, 1976.

the development of agricultural and forestry based processing industries; the decentralisation of office employment; town development in rural regions, and measures to co-ordinate approaches by regional and local development plans.

This last point on co-ordination is of vital importance because, as Slee (1981) has argued, funds for the modernisation of agriculture only throw agricultural employees out of work so that more public money, namely, regional planning funds, have then to be used to re-employ them elsewhere. This lack of co-ordination has also been attacked by Gilg (1980) who has advocated an Upland Resource Development Commission to integrate grants for the rural economy and to draw up regional resource development plans.

Unfortunately, policies for the rural economy still tend to be unco-ordinated and the agencies charged with implementing them fragmentary. Apart from regional planning policies, for example, Industrial Development Certificates, and grants and subsidies which are not specifically urban or rural, there are a number of specific rural agencies, as well as two national economic agencies, the Welsh and Scottish Development Agencies (Carney and Hudson, 1978, 1979). The oldest of these, the Development Commission and its offshoot COSIRA (Council for Small Industries in Rural Areas), had a considerable revival in the 1970s, following the Treasury report on rural depopulation. In the light of this report, the government asked the Development Commission to provide 1,500 extra rural jobs a year. In order to achieve this total the Commission has funded advance factories in key rural centres in areas suffering from depopulation, unemployment and ageing population, in collaboration with local councils who draw up both short-term and long-term plans for employment provision (Northfield, 1977). At a more local scale the Highlands and Islands Development Board and the Development Board for Rural Wales have performed similar tasks since 1965 and 1976 respectively (Grieve, 1973; Minay, 1977). Central to the work of all three is the provision of advice, and funds for social provision as well as financial help and the construction of advance factories. The net result of this work is that the County Planning Officer's Society (1980) was able to list 30 different schemes, merely as examples of the types of approach to reviving the rural economy and 15 agencies that could provide help or funds.

In summary, five main points have emerged from the above discussion and these broadly follow the conclusions reached by a Planning Exchange (1979) seminar on 'Population and Employment in Rural Areas':

1. There is a difference in the scale of development and aid necessary for regeneration in rural and urban areas. Rural regeneration can be much cheaper.
2. Key settlement strategies may be at odds with what is required locally to sustain communities. A balance between the costs of service provision and more scattered development may have to be struck.
3. A large-scale industrial development in a rural area does not necessarily generate additional industrial growth.
4. It is necessary to co-ordinate the activities of public bodies, especially those with an areal remit compared to those with a subject remit.
5. There is a need to encourage the establishment and growth of locally-based industries.

Although these conclusions were based on Scottish experience, similar findings have been reported from Ireland (Commins, 1978) and the inter-relationships between agricultural, industrial, human resource and social policies were once again stressed, as a further reminder that policies for rural population and employment are inextricably linked to policies for the wider world as well, but that policies for one period are not necessarily valid for another.

This has been emphasised most forcibly in the early 1980s, for since most of the work outlined in this chapter has been written Britain and most of the world has been plunged into the deepest recession since the 1930s, and according to some writers the recession is masking or foreshadowing a development towards a very different economy, in which automation and computers will render nearly all present work patterns redundant. Already service industries account for half of all rural employment and tertiary employment has been and is likely to be the only substantial growth area in the employment field, as shown in Table 4.9.

And yet very few of the policies and proposals outlined above have emphasised service employment, concentrating instead on manufacturing employment which with automation will surely go the way of agriculture, for the rises shown in Table 4.9 are mainly due to re-location of urban jobs in the countryside not new growth. Looking even further ahead, it is possible and even probable that a much smaller proportion of people's lives will be spent in paid employment, and people will retire earlier, work far fewer hours and days per week and will have very long holidays so that work can be shared round. The

Table 4.9: Trends in Employment by Sector for Selected Counties

Period	County	Primary	Manufacturing	Service	Total
1971-73	Cumbria	−7.2	+ 2.0	+ 4.5	+ 2.9
1971-76	N. Yorkshire	+ 7.3	+ 7.2	+ 11.9	+ 10.2
1971-75	Lincolnshire	−1.3	+ 0.7	+ 2.0	+ 1.2
1971-76	Cambridgeshire	− 13.7	+ 6.3	+ 23.6	+ 7.8
1971-75	Norfolk	− 11.1	+ 2.1	+ 18.3	+ 10.3
1971-74	Suffolk	−5.3	+ 8.2	+ 12.6	+ 9.6
1971-75	Cornwall	− 11.2	−0.3	+ 13.8	+ 7.0
1971-75	Shropshire	− 14.7	+ 3.7	+ 12.3	+ 6.6
1976	Rural areas	12	39	49	100
1976	England	2.5	42	55.5	100

Source: National Council of Voluntary Organisations, 1980.

effect on settlement patterns could be profound and Blake (1979) has suggested: 'The constrictions of our older cities will become more and more irksome. There will be a demand for more second homes in the countryside, and some families may reverse the process altogether by moving to the country and keeping a town pad for use while at work.' In other scenarios, almost everyone could move to low density countryside houses, and communicate via electronic technology that is already available, or alternatively people could recolonise the countryside and run smallholdings which on some estimates produce far more food per unit of energy used than technologically dominated and energy inefficient agribusiness (Anderson, 1975). If these forecasts are correct the population changes revealed by the 1981 census, as shown in Figure 4.1, are only the precursors of a massive recolonisation of the countryside during the rest of the century, but they ignore a number of issues, mainly the immense social capital invested in our cities, the loss of landscape quality that could occur, and unknown effects on food production. It is, however, very likely that the rural population geography of the next few years will become increasingly concerned with where people want to live and how they want to occupy their time rather than with the concerns outlined in this chapter which have been where people have been forced to live and be employed by factors outside their control.

References

Anderson, M.A. (1975), 'Land Planning Implications of Increased Food Supplies', *The Planner*, 61, 381-3

Association of County Councils (1979), *Rural Deprivation* (The Association, London)

Association of District Councils (1978), *Rural Recovery: A Strategy for Survival* (The Association, London)

Beaujeu-Garnier, J. (1978), *Geography of Population* (Longmans, London)

Best, R.H. and Rogers, A.W. (1973), *The Urban Countryside* (Faber, London)

Best, R.H., Jones, A.R. and Rogers, A.W. (1974), 'The Density Size Rule', *Urban Studies*, 11, 201-8

Blake, J. (1979), 'Job Prospects', *Town and Country Planning*, 47, 7-10

Bracey, H. (1958), 'Some Aspects of Rural Population in the United Kingdom', *Rural Sociology*, 23, 385-91

Buksmann, P. (1980), 'An Approach to Agricultural Migration in Bolivia' in *The Geographical Impact of Migration*, edited by White, P. and Woods, R., (Longmans, London), 108-28

Carney, J.G. and Hudson, R. (1978), 'The Scottish Development Agency', *Town and Country Planning*, 45, 507-9

Carney, J.G. and Hudson, R. (1979), 'The Welsh Development Agency', *Town and Country Planning*, 48, 15-16

Census Research Unit (1980), *People in Britain: A Census Atlas* (HMSO, London)

Champion, A.G. (1973), 'Population Trends in England and Wales', *Town and Country Planning*, 41, 504-9

Champion, A.G. (1976), 'Evolving Patterns of Population Distribution in England and Wales 1951-71', *Transactions of the Institute of British Geographers. New Series*, 1, 401-20

Clark, C. (1977), *Population Growth and Land Use* (Macmillan, Basingstoke)

Cloke, P.J. (1978), 'Changing Patterns of Urbanisation in Rural Areas of England and Wales', *Regional Studies*, 12, 603-17

Commins, P. (1978), 'Socio-economic Adjustments to Rural Depopulation', *Regional Studies*, 12, 79-94

Council of Europe (1976), *On the Steps which can be Taken to Reduce Depopulation of Rural Regions: Council Resolution ? (76)* 26 (HMSO, London)

Country Landowners Association (1980), *Employment in Rural Areas* (The Association, London)

County Planning Officer's Society (1980), *The Rural Economy: A Handbook of Ideas* (The Society, Brecon)

Craig, J. (1972), 'Population Potential and Population Density', *Area*, 4, 10-12

Craig, J. (1974), 'How Arbitrary is Population Potential', *Area*, 6, 44-6

Craig, J. (1979), 'Population Density: Changes and Patterns', *Population Trends*, 17, 12-16

Craig, J. and Frosztega, J. (1976), 'The Distribution of Population in Great Britain by Ward and Parish Density, 1931, 1951 and 1961', *Area*, 8, 187-90

Drudy, P.J. (1978), 'Depopulation in a Prosperous Agricultural Sub-region', *Regional Studies*, 12, 49-60

Drudy, P.J. and Drudy, S.M. (1979), 'Population Mobility and Labour Supply in Rural Regions: North Norfolk and Galway', *Regional Studies*, 13, 91-9

Drudy, S.M. (1975), 'The Occupational Aspirations of Rural School Leavers', *Social Studies*, 4, 230-41

Edwards, J.A. (1971), 'The Viability of Lower Size-order Settlements in Rural Areas: The Case of North-east England', *Sociologia Ruralis*, 11, 247-75

Edwards, J.A. (1973), 'Rural Migration in England and Wales', *The Planner*, 59, 450-3

Ehrlich, P.R. and Ehrlich, A.H. (1972), *Population Resources Environment: Issues in Human Ecology* (W.H. Freeman, London)

Gasson, R. (1969), 'Occupational Immobility of Small Farmers', *Journal of Agricultural Economics*, 20, 279-88

Gilg, A.W. (1976), 'Rural Employment' in *Rural Planning Problems*, ed. G.E. Cherry (Leonard Hill, Glasgow), 125-72

Gilg, A.W. (1980), 'Planning for Rural Employment in a Changed Economy', *The Planner*, 66, 91-3

Gilg, A.W. (ed.) (1981), *Countryside Planning Yearbook: Volume Two* (Geo Books, Norwich)

Grieve, R. (1973), 'Scotland: Highland Experience of Regional Government', *Town and Country Planning*, 41, 172-5

Grigg, D.B. (1976), 'The World's Agricultural Labour Force 1800-1970', *Geography*, 61, 194-202

Grigg, D.B. (1980), 'Migration and Overpopulation' in *The Geographical Impact of Migration*, ed. White, P. and Woods, R. (Longmans, London), 60-83

Grigg, D.B. (1981), 'The Historiography of Hunger: Changing Views on the World Food Problem', *Transactions of the Institute of British Geographers, New Series*, 6, 279-92

Hakim, C. (1978), *Census Data and Analysis: A Selected Bibliography* (Office of Population Censuses and Surveys, London)

Hall, P., Drewett, R., Gracey, H. and Thomas, R. (1973), *The Containment of Urban England* (Allen and Unwin, London)

Hampshire County Council (1966), *Village Life in Hampshire* (The Council, Winchester)

Hannan, D.F. (1969), 'Migration Motives and Migration Differentials Among Rural Youth', *Sociologia Ruralis*, 9, 195-220

Hodge, I. and Whitby, M. (1981), *Rural Employment* (Methuen, London)

Jackson, V.J. (1968), *Population in the Countryside: Growth and Stagnation in the Cotswolds* (Cass, London)

Johnston, R.J. (1965), 'Components of Rural Population Change', *Town Planning Review*, 36, 279-93

Johnston, R.J. (1967), 'A Reconnaissance Study of Population Change in Nidderdale 1951-61', *Transactions of the Institute of British Geographers*, 41, 113-23

Johnston, R.J. (1971), 'Resistance to Migration and the Mover-stayer Dichotomy: Aspects of Kinship and Population Change in an English Rural Area', *Geografiska Annaler*, 538, 16-27

Jones, H.R. (1965), 'A Study of Rural Migration in Central Wales', *Transactions of the Institute of British Geographers*, 37, 31-45

Jones, H.R. (1976), 'The Structure of the Migration Process: Findings from a Growth Point in Mid-Wales', *Transactions of the Institute of British Geographers, New Series*, 1, 421-32

Jones, H.R. (1981), *Population Geography* (Harper and Row, London)

Kennett, S. and Spence, N. (1979), 'British Population Trends in the 1970s', *Town and Country Planning*, 48, 220-3

Kirby, A.M. and Tarn, D. (1976), 'Some Problems of Mapping the 1971 Census by Computer', *Environment and Planning A*, 8, 507-13

Lewis, G.J. (1979), *Rural Communities* (David and Charles, Newton Abbot)

Lonsdale, R. and Holmes, J. (eds.) (1981), *Settlement Systems in Sparsely Settled Regions* (Pergamon, Oxford)

Mabogunje, A. (1970), 'Systems Approach to a Theory of Rural-urban Migration',

Geographical Analysis, 2, 1-15

McCallum, J.D. and Adams, J.G.L. (1981), 'Employment and Unemployment Statistics for Rural Areas', *Town Planning Review*, 52, 157-66

Minay, C.L.W. (1977), 'A New Rural Development Agency', *Town and Country Planning*, 45, 439-43

National Council for Voluntary Organisations (1981), *Jobs in the Countryside* (The Council, London)

Northfield, Lord (1977), 'The Role of the Development Commission', *Town and Country Planning*, 45, 304-7

Office of Population Censuses and Surveys (1978), *Socio-economic Classification of Local Authority Areas* (HMSO, London)

Office of Population Censuses and Surveys (1980), *Population Density and Concentration in Great Britain, 1951, 1961 and 1971* (HMSO, London)

Office of Population Censuses and Surveys (1981), *Census 1981: Preliminary Report* (HMSO, London)

Ogden, P. (1980), 'Migration, Marriage and the Collapse of Traditional Peasant Society' in *The Geographical Impact of Migration*, ed. White, P. and Woods, R. (Longmans, London), 152-79

Openshaw, S., Cullingford, A. and Gillard, A. (1980), 'A Critique of the National Classifications of OPCS/PRAG', *Town Planning Review*, 51, 421-39

O'Riordan, T. (1981), *Environmentalism* (Pion, London)

Pacione, M. (1980), 'Quality of Life in a Metropolitan Village', *Transactions of the Institute of British Geographers' New Series*, 5, 185-206

Pacione, M. (1981), *Problems and Planning in Third World Cities* (Croom Helm, London)

Planning Exchange (1979), *Population and Employment in Rural Areas* (The Exchange, Glasgow)

Pressat, R. (1978), *Statistical Demography* (Methuen, London)

Ravenstein, E.G. (1885), 'The Laws of Migration', *Journal of the Royal Statistical Society*, 48, 167-235

Rees, P.H. and Wilson, A.G. (1975), 'A Comparison of Available Methods of Population Change', *Regional Studies*, 9, 39-61

Rees, P.H. and Wilson, A.G. (1976), *Spatial Population Analysis* (Arnold, London)

Registrar General, Scotland (1981), *Census 1981: Preliminary Report* (HMSO, London)

Rhind, D. (1975), 'Mapping the 1971 Census by Computer', *Population Trends*, 2, 9-12

Rhind, D. (1976), 'People Mapped by Laser Beam', *Geographical Magazine*, 49, 148-52

Robertson, I.M.L. (1976), 'Accessibility to Services in the Argyll District of Strathclyde: A Locational Model', *Regional Studies*, 10, 89-95

Saville, J. (1957), *Rural Depopulation in England and Wales 1851-1951* (Routledge and Kegan Paul, London)

Shaw, M.J. (1979), *Rural Deprivation and Planning* (Geo Books, Norwich)

Slee, R.W. (1981), 'Agricultural Policy and Remote Rural Areas', *Journal of Agricultural Economics*, 32, 113-21

Smith, C.T. (1951), 'The Movement of Population in England and Wales 1851-1951', *Geographical Journal*, 117, 200-10

Treasury, Her Majesty's (1976), *Rural Depopulation: Report by an Inter-departmental Study Group* (The Treasury, London)

Turnock, D. (1968), 'Depopulation in North-East Scotland with Reference to the Countryside', *Scottish Geographical Magazine*, 84, 256-68

Vince, S.W.E. (1952), 'Reflections on the Structure and Distribution of Rural

Population in England and Wales 1921-31', *Transactions of the Institute of British Geographers*, 18, 53-76

Webber, R.J. (1980), 'A Response to the Critique of the National Classification of OPCS/PRAG', *Town Planning Review*, 51, 440-50

White, P. (1980), 'Migration Loss and the Residual Community: a Study in Rural France' in *The Geographical Impact of Migration*, eds. White, P. and Woods, R. (Longmans, London), 198-222

White, P. and Woods, R. (eds.) (1980), *The Geographical Impact of Migration* (Longmans, London)

Willatts, G.C. and Newson, M.G. (1953), 'The Geographical Pattern of Population Changes in England and Wales 1921-51', *Geographical Journal*, 119, 431-54

Woodruffe, T. (1976), *Rural Settlement Policies* (Oxford University Press, Oxford)

Zelinsky, W. (1970), *A Prologue to Population Geography* (Prentice-Hall, New York)

5 HOUSING

A.W. Rogers

Housing dominates the land-use structure of the built countryside, yet it has frequently been ignored by rural geographers in favour of other topics. Even when geographers have studied rural housing it has seldom been in a comprehensive way but has rather focused on a single topic, such as second homes or agricultural housing. The part played by housing in the rural economies of advanced societies has only recently returned as a topic of real concern to geographers.

This was not always the case, however. In the past rural housing was considered by many geographers to be an essential element in their understanding of regional geography. In Europe, and to some extent in North America, the distinctive styles of housing in the countryside were considered to provide a direct link both with the agricultural economy of the region and with its cultural geography. An area of study grew up which related vernacular architectural style and housing plan to the economic and social functions which housing had to perform and to the process of settlement by different peoples. This tradition within rural geography was particularly strong in France during the early years of the twentieth century (eg, Brunhes, 1920; Demangeon, 1920) but it can still be found today. An example of this school would be Houston's (1964) study of house types in the western Mediterranean where much emphasis is placed upon the link between house design and the related agricultural economy. Houston recognised the correlation between multiple-storey housing and pastoral economies (where animals live on the ground floor with people dwelling above) and single-storey dwellings which are more characteristic of areas of cereal cultivation. Other examples of this type would include Birch's (1974) study of farm housing in the Corn Belt which followed on from Trewartha's (1948) earlier research in the United States. In Britain, the studies of farm housing by the landscape historian Hoskins (1970) are from the same school.

This relative neglect by geographers of rural housing as a topic of enquiry, or at least its historical bias towards rather descriptive cultural geography, means that the student has frequently to depend upon the work of other academic disciplines, notably economics and sociology,

106

if he is to build up a comprehensive picture of rural housing in advanced societies. This cross-referencing has become all the more necessary as the economies of advanced capitalist countries have changed, particularly during the twentieth century, and especially as their rural and urban components have become inextricably mixed. Rural housing in such countries has increasingly become divorced from the agricultural economy to which it was previously tied. While, as will be seen, rural housing is still an important component in agricultural policy, its significance has widened to reflect the increasing desires and aspirations of a rich and mobile urban society.

An important key, then, to the understanding of the part which housing plays within rural geography lies in the social and economic changes which have transformed the rural areas of the Western world. In particular it is the demographic changes that are now being seen in the second half of the twentieth century which are forcing a reassessment of the role of rural housing. Most Western countries have, during the last decade or so, seen a substantial trend towards 'counter-urbanisation' or, as the Americans call it, the 'population turnaround' (Brown and Wardell, 1980). The invasion of the countryside from the towns has a profound effect upon the state of rural housing.

Rural housing in advanced societies, therefore, has now to perform several functions and not just the one of sheltering the agricultural population. Particularly since the Second World War housing in the countryside has increasingly been taken over by commuters who, while still working in urban areas, have been able to move their households to a more pleasant environment in the countryside. Others have gone further and have, at least in part, moved their occupations also.

The declining relative significance of agricultural employment in all Western societies has combined with a broadening of the rural employment structure and the introduction of new manufacturing and service-based industries (Beale, 1980). Many people, too, have chosen to use the rural home as the basis of a growing recreational experience. Finally, all advanced societies, characterised by an ageing population structure, have seen the movement of growing numbers of older people to live in rural areas.

The significant point is that rural housing now has to perform all these functions and not just one. Moreover, these roles stretch across all the functions of our modern mixed economy and, in so doing, range from the provision of basic shelter for a residual, often relatively poor, agricultural population, to the most extreme forms of conspicuous and inessential consumption for a wealthy, urban-oriented middle class. As a result the geography of rural housing has changed from being an integral part of a rural, generally agricultural economy, and has become as involved and multi-faceted as its urban counterpart. In the process it has inevitably become concerned with issues of social justice and equity, state involvement and private interests. It is these focal points around which much of this chapter revolves.

The Structure of Rural Housing

The trend in Western societies towards counter-urbanisation is, however, at least for the majority of the population, a fairly recent one. In Britain, for example, the first signs of a demographic shift towards rural areas was really only apparent in the 1960s and it was not until the early 1980s that it really impinged upon public awareness (Office of Population Censuses and Surveys, 1981). Similarly in the United States, while the evidence for population growth in some non-metropolitan areas was discernible in the 1960s, it was not until the mid-1970s that it was generally acknowledged (Beale, 1977).

The structure of housing in the rural areas of Western societies is, on the other hand, a very old one and even in the most urbanised and advanced society it still reflects the rural economy from which it originated. This means that rural housing is on the whole generally older and in much poorer condition than its urban equivalent and it is inhabited by households which generally have lower incomes than their counterparts in the towns and cities.

A major characteristic of rural housing systems is their tenurial structure, which is invariably dominated by owner occupation. The norm in many rural areas, both of Western Europe and North America, is still the self-employed entrepreneur and his family living in their own property. Even in countries such as England where a farm tenancy system is of some antiquity, the owner-occupier has always been dominant and, indeed, has increased substantially in importance in recent years (Rogers, 1971). In 1971 some 54.5 per cent of households in English rural districts owned their own houses (Office of Population Censuses and Surveys, 1981). The proportion of owners is generally much higher in other countries; in the United States, for example, the proportion of owner-occupiers rises to 78 per cent (Limmer and McGough, 1979), though this figure increased from around 60 per cent in 1950 (Bird and Kampe, 1977).

Table 5.1: The Age of Rural Housing, France, 1975

	Before 1915	1915–48	1949–61	1962–67	1968+	Total
	%	%	%	%	%	%
Ordinary dwellings in rural communes	62	11	7	6	14	100
Ordinary dwellings in urban areas (excluding Greater Paris)	31	16	18	15	20	100
Ordinary dwellings in Greater Paris	30	23	17	13	17	100
Total ordinary dwellings	39	16	15	12	18	100

Source: *Recensement général de la population de 1975*, quoted by Fennell, 1981.

By contrast, rented property in rural areas is generally restricted to a minority of the population and a minority which is frequently disadvantaged in some way and cannot therefore afford to own. In Britain the rented house is particularly important for hired farm workers where the tenancy of the property is 'tied' to employment on the farm. Over half of all agricultural employees in England and Wales now live in tied accommodation (Gasson, 1975; Irving and Hilgendorf, 1975) and the proportion rises to three-quarters in Scotland where labour must frequently be housed on isolated holdings (Mackenzie and Martin, 1975).

Relative to urban housing, housing in rural areas is generally older and in poorer condition. The twin factors of the age of the housing stock coupled with the low incomes of rural dwellers (particularly those involved in agriculture) mean that even in advanced societies rural dwellers have fewer housing amenities. In the old economies of Western Europe most of the rural housing pre-dates the twentieth century. A good example is France (Table 5.1), where more than 60 per cent of all the homes in rural communes date from before the Great War, whereas only half this proportion is found in the urban areas. The proportion of pre-1915 houses rises to over 70 per cent in the case of those houses belonging to farmers and farm workers. Very similar figures can be quoted for other countries in Western Europe — 59 per cent of farm dwellings in Belgium in 1970 were built before 1919 and the corresponding figure for West Germany (1968) was 68

Table 5.2: Rural Housing Conditions and Amenities, France, 1975

	No inside flush toilet		No bath/shower		No central heating		Inadequate water supply	
	Principle residence with 1 dwelling	Farms	Principle residence with 1 dwelling	Farms	Principle residence with 1 dwelling	Farms	Principle residence with 1 dwelling	Farms
	%	%	%	%	%	%	%	%
Rural communes	41	63	43	58	65	82	6	10
Urban areas (excluding Paris)	28	55	27	49	48	76	2	9
Greater Paris	14	25	18	20	23	32	2	2
Total	32	62	32	57	52	81	3	9

Note: Principle residences exclude second homes.

Source: As for Table 5.1.

Table 5.3: Contrasts in Rural Housing Amenities, France, 1975

Western regions	%	Eastern regions	%
Haute-Normandie	23	Lorraine	65
Bretagne	21	Alsace	55
Poitou-Charente	12	Franche-Comté	56
Aquitaine	16	Rhône-Alpes	45
Midi-Pyrénées	28	Languedoc-Roussillon	73

Note: Percentages refer to population of rural communes whose houses are connected to the public sewer.

Source: As for Table 5.1.

per cent. Rural housing in the United States, however, shows a rather younger age profile than in other advanced societies. In 1976 two-thirds of all rural housing had been built since 1939. In part this reflects the much younger history of the nation but an important subsidiary factor is the relatively recent development of mobile homes, especially since the mid-1960s (Mikesell, 1973). In 1976 more than one rural house in ten in the USA was a mobile home (Limmer and McGough, 1979), as compared with only one house in twenty in the country as a whole.

The problems of rural housing conditions can also be illustrated with reference to France (Table 5.2), where the majority of houses in the rural communes still lack basic amenities such as inside toilets and baths. Despite substantial investment from the national government and from European Community sources, French rural housing, particularly within the agricultural sector, remains in poor condition. And yet even this represents an improvement on conditions at mid-century. Even as late as the 1950s there were still large areas of rural France, particularly in the western regions of Brittany, Poitou-Charente and Aquitaine, where animals lived in the same buildings as the farmer and his family, and where farm workers were accommodated in barns and stables (Pitie, 1969). This distinction between the relatively better-housed east and the poorer conditions of the west in France is still evident today as indicated, for example, by the proportion of the rural population whose houses are connected to a public sewer (Table 5.3).

This picture of poor-quality rural housing, particularly agricultural housing, and the continued existence of rural regions where conditions are particularly bad, can be repeated for all the nations of the Western world. In England, for example, the 1976 *House Condition Survey* showed that a higher proportion of rural houses were judged unfit than

Table 5.4: Housing Conditions and Amenities, England, 1976

	Conurbations	Other urban areas	Rural areas	England
	thousands (% all dwellings in area)			
Unfit dwellings	315	255	224	794
	(4.8)	(3.9)	(5.6)	(4.6)
All dwellings	6,598	6,534	3,983	17,115
	(100.0)	(100.0)	(100.0)	(100.0)
Lack of amenities				
WC outside	456	397	230	1,083
	(6.9)	(6.1)	(5.8)	(6.3)
Fixed bath in bathroom	347	273	180	800
	(5.3)	(4.2)	(4.5)	(4.7)
Wash basin	455	345	191	991
	(6.9)	(5.3)	(4.8)	(5.8)
Sink	6	13	24	43
	(0.1)	(0.2)	(0.6)	(0.3)
Hot and cold water at 3 points	525	417	231	1,173
	(8.0)	(6.4)	(5.8)	(6.9)
One or more of the basic amenities	677	529	287	1,493
	(10.3)	(8.1)	(7.2)	(8.7)
All dwellings with or without ameniti	6,598	6,534	3,983	17,115
	(100.0)	(100.0)	(100.0)	(100.0)

Source: Department of the Environment, 1978.

in the conurbations (Table 5.4). And yet as a result both of public and private investment there have been major improvements in recent years. The number of unfit houses declined by 12 per cent between 1971 and 1976 and by no less than 46 per cent between 1967 and 1971. The more rural and less accessible areas of the country, where rural incomes are at their lowest, are especially badly affected. In particular East Anglia, the South West of England and the Welsh borderland remain areas of bad rural housing. In some parts of Herefordshire more than one house in five lacks basic amenities and one in six are unfit by modern standards (Hereford and Worcester County Council, 1978).

Equally, in the United States massive improvement in housing conditions have not benefited some rural areas (USDA, 1971). In the nation as a whole there was a dramatic improvement in the quality of rural housing after the Second World War. In 1950 some 9.08 million rural houses were considered substandard, of which the majority (56.4 per cent) was in in the owner-occupied sector. By 1975 this figure had fallen to 1.92 million properties and the majority of these (51.7 per cent) were now in the rented sector (Bird and Kampe, 1977). Even so, substandard rural housing still accounts for just over half of all substandard housing in the United States even though only 28 per cent of all households live in the rural areas. Some regions, notably in the South and in Appalachia, still exhibit particularly bad conditions. The Central Appalachian Region typifies this residual problem (Deaton and Hanrahan, 1973). Even as late as 1960 no less than 43.8 per cent of all housing in the sixty Counties which make up the region (parts of Kentucky, Virginia, West Virginia and Tennessee) were officially considered as deteriorating or dilapidated. Even after substantial improvements in the 1960s the general standard, as judged for example by availability of satisfactory plumbing facilities, was still poor, particularly in parts of Kentucky (Table 5.5).

Table 5.5: Housing Amenities in the Central Appalachian Region, United States 1970

Region	Number of housing units	Housing units lacking full plumbing facilities	
		No.	%
Kentucky (26 counties)	163,444	74,608	45.5
Tennessee (18 counties)	103,217	33,083	32.2
Virginia (17 counties)	61,402	24,780	40.2
West Virginia (9 counties)	119,856	32,164	26.7
Central Appalachian region (60 counties)	447,919	164,635	36.7
Four state total	4,433,892	721,184	16.3
United States total	68,418,062	5,168,646	7.6

Source: US Department of Commerce, Bureau of the Census, *Census of Housing*, 1970.

Rural Housing, Economy and Society

These contrasts in housing quality and facilities, both between the rural and the urban sectors and within the rural areas of Western societies, do not of course exist in a vacuum. The nature of rural housing is intimately bound up with the complex of social and economic structures which make up national and regional economies. The spatial pattern of good and bad housing, of ownership and tenancy and of state housing aid is not a random one but rather reflects the economic health of different rural areas and the attitudes of society to its constituent members made on the basis of income, occupation, class and race. Differentials and discrimination exist in rural areas just as much as they do in the more widely researched urban sector.

Two studies, one for England and the other for France, which have attempted to look at the picture in the round provide a useful starting point. Both emphasise this complex of economic and social factors which determines the nature of rural housing. Other studies can then illustrate particular points.

A major study of English rural housing carried out by the present author and colleagues (Dunn, Rawson and Rogers, 1981) has especially stressed the socio-economic environment in which rural housing is found. Housing tenure, for example, provides a clear instance of social and economic polarisation. Privately rented housing is strongly concentrated within the agricultural sector, both for farmers and for farmworkers, whereas housing provided by the local authority is commonly used by other manual workers. At the other extreme, the owner-occupied sector is very much the preserve of the professional and non-manual worker. Housing quality also is strongly linked with occupation and social class, with higher quality housing being used by the middle classes and the poorer housing used by manual workers, particularly those in agriculture.

The complex of social and economic factors alluded to above has a distinct spatial dimension. In England the central urban core of the country stretching north-westwards from London and the south east towards Manchester and Liverpool is reflected in the nature of the housing in the associated countryside. Away from this core, in the peripheral rural regions where agriculture is more dominant, rural housing is generally poorer in quality, older in age and inhabited by less-advantaged households. This pattern is clearly seen from Figure 5.1, which illustrates the distribution of the seven basic 'rural housing profiles' recognised in a cluster analysis of English rural housing in 1971

**Figure 5.1: Rural Housing Profiles 1971, England
for Key see Table 5.6**

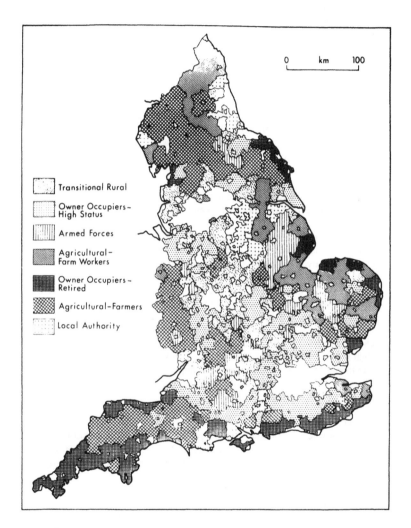

(Dunn, Rawson and Rogers, 1981). Table 5.6 provides brief descriptions of these seven profiles. The more truly rural areas of the North of England, the eastern counties, the Welsh Borders and the South West on the one hand, are particularly associated with profiles 1, 2 and 3, that is where agricultural employment is strong or where retirement, either of the indigenous population or because of immigration, is an

Table 5.6: The Character of Rural Housing Profiles, England, 1971

1. *Agricultural — farmworkers (42 cases)*
   A high proportion of agricultural workers and low proportions of professional and non-manual workers. Above average proportion of private rented unfurnished tenancies and also a greater than average number of households lacking exclusive use of one or more basic amenities. High unemployment rates. Low level of car ownership.

2. *Agricultural — farmers (91 cases)*
   A predominantly agricultural group with low population totals and high proportions of farmers and agricultural workers. Many households living in rented unfurnished accommodation. Low occupancy rates. Above average proportion of pensioners.

3. *Owner-occupiers — retired (42 cases)*
   Large retired population. Low economic activity rates; high unemployment rates. High level of owner occupation and under-occupancy of dwellings.

4. *Transitional rural (90 cases)*
   A large profile incorporating a variety of characteristics. High female economic activity rate. Above average proportion of non-manual workers; slightly more than average percentage of children. Fewer pensioners and agricultural workers than the average for rural districts. Renting privately unfurnished is a smaller than average tenure category and there is less under-occupancy.

5. *Owner-occupiers — high status (88 cases)*
   High proportion of heads of household employed in professional and non-manual occupations and correspondingly above average levels of education. Two car ownership and owner-occupation are both above average.

6. *Armed forces (36 cases)*
   A small profile dominated by armed forces personnel and households living in rented furnished accommodation. Above average proportion of children leading to more overcrowding. High male economic activity rate. Few pensioners.

7. *Local Authority housing (26 cases)*
   A small group of rural districts with high proportions of households living in local authority rented accommodation. Skilled manual workers dominate the occupational structure. Relatively few households own cars. Some overcrowding of dwellings.

Source: Dunn, Rawson and Rogers, 1981.

important function of the rural housing stock. In contrast, in the 'conurban' regions (Green, 1971), profiles 4, 5, 6 and 7 are dominant, that is where incomes are generally higher, employment opportunities are more varied, housing quality is better, the demographic structure is more vibrant or where state involvement in housing provision has been particularly effective.

A rather similar form of multivariate analysis, using a principal components technique, has been applied by Ogden (1978) in a study of the eastern Massif Central in France. In defining the main components differentiating the region, the analysis stressed particularly the significance of poor housing quality and the links between rural housing and the social and economic structure of the region. Thus areas where housing generally lacked amenities such as internal toilets, baths and running water were also found to have a high proportion of old people, a tradition of population outmigration and a dominance of agricultural employment. Moreover, such areas were generally to be found at higher altitudes and in less accessible parts of the region.

These national or regional studies provide an indication at an aggregate level of the social and spatial differences which can be seen in rural housing in Western societies. It is possible to look at some of the specific factors which provide this aggregate view in more detail. Three factors have been shown to be particularly important − household income, occupation and, notably in the United States, minority groups.

Housing and Income

Income naturally provides a major determinant both of housing quality and of housing choice. Lower income households tend to inhabit rented accommodation more frequently than owner-occupied property, but in whichever tenure sector they are found they are much more likely to suffer from poorer housing conditions. Moreover, other attributes of their life-styles are likely to make them deprived relative to richer households. Case studies from the English shires clearly illustrate this relationship. Table 5.7 presents data for two rural areas − Cotswold District (Gloucestershire) and South Oxfordshire District (Oxfordshire) − which show clearly how income determines housing tenancy, housing quality and lifestyle.

Evidence from other countries supports this relationship. Table 5.8 presents data for the non-metropolitan areas of the United States. As noted earlier, there has since 1950 been a very substantial reduction in the numbers of substandard rural houses. Equally there has been a substantial improvement in the level of rural incomes, and this factor,

Table 5.7: Rural Housing and Rural Incomes, England, 1977

Gross annual income	Owner-occupiers %	Local authority tenants %	Private unfurnished tenants %	Private furnished tenants %	Tied housing tenants %	Households lacking amenities %	Households lacking a car %
South Oxfordshire District							
less than £3,000	22.6	50.0	68.6	22.0	31.5	68.4	66.0
£3,000–£6,000	37.3	39.3	22.9	55.6	48.1	31.6	30.9
more than £6,000	40.1	10.7	8.5	22.2	20.4	0.0	3.1
	100.0	100.0	100.0	100.0	100.0	100.0	100.0
Cotswold District							
less than £3,000	35.2	63.9	59.4	63.6	60.3	76.9	84.0
£3,000–£6,000	32.4	33.3	31.3	27.3	33.7	23.1	16.0
more than £6,000	32.4	2.8	9.3	9.1	6.0	0.0	0.0
	100.0	100.0	100.0	100.0	100.0	100.0	100.0

Source: Dunn, Rawson and Rogers, 1981.

Table 5.8: Rural Housing and Rural Incomes, United States, 1950 and 1975

Type of units and household income	1950		1975	
	000s	%	000s	%
Total units				
less than $ 2,000	7,105	45.9	5,189	22.7
$ 2,000-$ 3,999	5,338	34.5	5,478	24.0
$ 4,000-$ 5,999	1,996	12.9	4,497	19.7
$ 6,000 and over	1,042	6.7	7,693	33.6
Total	15,481	100.0	22,857	100.0
Substandard units as % of total units				
less than $ 2,000		78.5		21.1
$ 2,000-$ 3,999		50.6		9.5
$ 4,000-$ 5,999		29.3		4.2
$ 6,000 and over		20.6		1.6
Total		58.7		8.4

Note: Household incomes adjusted to 1950 values.

Source: US Census of Housing and Population, 1950; *Annual Housing Survey, 1975.*

coupled with Federal, State and local housing programmes, has obviously been a major cause for the overall reduction in poor housing. However, it is equally evident that poor housing is still very much concentrated in the low income sector. Although in 1975 only one in five rural households with incomes below $2,000 lived in substandard housing as compared with four out of five households in 1950, the reduction in poor housing for the higher income groups had been relatively higher. Substandard housing for the richer households (earning in excess of $6,000) had declined tenfold since 1950, whereas that for households earning less than $2,000 had dropped by a factor of four.

One aspect which is related to this situation is the higher cost of rural housing in relation to rural incomes. Because services are more expensive to provide in rural areas and because scale economies in construction are less common, housing is invariably more expensive to purchase (Clark, 1980; Dunn, Rawson and Rogers, 1981). Moreover, since rural incomes are invariably lower than in urban areas and decline with increasing 'rurality' (Whitby and Hubbard, 1982), the rural dweller is doubly disadvantaged. This has been demonstrated with regard to rural housing in the United States (Limmer and McGough, 1979) where the evidence for house prices and incomes shows that the cost of

Table 5.9: Rural Housing Costs and Rural Incomes, United States, 1975

Ratio of adequate housing cost to income	All United States households %	United States rural households %
Under 10 per cent	44.0	30.3
Under 20 per cent	74.3	65.6
Under 25 per cent	80.3	74.0
Under 30 per cent	84.4	78.9
Under 35 per cent	87.5	82.4
Under 40 per cent	89.9	85.1
Under 50 per cent	92.9	88.9
Under 60 per cent	94.9	91.2
Under 70 per cent	96.0	92.7

Source: Limmer and McGough, 1979.

adequate housing falls more heavily on rural families. Table 5.9 expresses housing cost relative to incomes and indicates that, at any point on the ratio scale, fewer rural households are able to secure adequate housing for a given income level. In other words, rural households must in general either spend a greater proportion of their income to find housing or must alternatively put up with substandard accommodation.

Housing and Occupation

The importance of the links between rural housing condition and employment have already been mentioned. In particular it was noted that in all Western societies there is a clear dichotomy which exists between agricultural and non-agricultural rural housing which is to the disadvantage of the agricultural sector. This point is further brought out by Table 5.10 in relation to French rural housing. Farmers and agricultural workers, both working and retired, have very much worse housing than other groups, notably the professional and middle classes. This picture is repeated elsewhere; in the United States, for example, the condition of farm housing is in general half as bad again as other rural housing.

This relationship between poor housing and agricultural employment is more than simply an interesting statistic. It has for long been an integral part of rural policy in many countries. Particularly within the European Community, poor rural housing has been equated with low agricultural efficiency and with rural depopulation, although policy

Table 5.10: Rural Housing Amenities and Employment in French Rural Communes, 1975

| Occupation | Households lacking: | | | | House- |
	Running water	Flush toilet	Bath/ shower	Central heating	holds *with* four basic facilities
Farmers	8	56	50	76	19
Agricultural workers	11	55	54	83	12
Industrialists/wholesalers	—	12	8	23	71
Tradesmen, fishing skippers and retailers	2	28	24	52	42
Liberal professions	1	7	4	13	84
Higher management	1	7	5	10	78
Middle management	1	15	12	37	59
Employees	2	26	25	54	42
Foremen, skilled workers and apprentices	2	28	28	57	39
Other primary and secondary sector workers	5	42	42	71	25
Service industry workers	5	40	40	63	31
Artists, clergy, army and police	2	22	22	53	43
Retired: non-agricultural	7	49	55	73	20
Retired: agricultural	14	71	77	86	8

Source: As for Table 5.1.

measures to date have tended to concentrate on income support through the Common Agricultural Policy rather than tackling social and economic issues such as housing in a more direct fashion. None the less there has been growing concern that poor living conditions not only contribute to a depressed agriculture but also siphon off investment which might otherwise be spent more productively in agriculture.

Housing and Minority Groups

The social and economic discrimination and its spatial counterpart of segregation and ghetto development which is common in the cities of Western nations has in general no obvious parallel in rural areas. The cultural clashes which exist for example in rural Wales over the question of second homes are insignificant when compared to the racial issues of the inner cities. Only in the United States is it really possible to find an equivalent where, particularly in the South, there is a significant black rural population.

Rural blacks suffer from the worst housing conditions of all groups in American society. While nationally one in five black households live in housing which has at least one physical deficiency, rural black households are affected at twice this rate. At least a third of rural black housing suffers from multiple flaws and this group far exceeds the second most poorly housed minority, Puerto Ricans, who are almost entirely resident in urban areas.

While low income levels provide the major explanation for the poor housing conditions of rural blacks, there is also evidence that this group suffers from inadequate housing because there exists discrimination in housing access. A study of rural housing discrimination by Marantz, Case and Leonard (1976) has argued that such discrimination has become institutionalised in so far as two separate markets for housing have grown up which have a built-in price differential which is to the disadvantage of black households. Their evidence from case studies in Arkansas, Georgia, North Carolina and Tennessee showed, moreover, that there is clear segregation of black households in small rural towns.

Rural Housing and the Leisure Society

In addition to the incursions of richer urbanites to live in rural areas, rural housing has also been at the centre of the growth in leisure and outdoor recreation. The second home in the countryside has become a significant element in the rural scene of virtually all Western societies. For many households the ownership of a second home has provided a recreational facet to the separation of home and work place and it is in this context that Clout (1974) has suggested that second homes be regarded as a form of 'seasonal suburbanisation'.

The first attempt by a geographer to provide a systematic study of second homes was Wolfe's account of summer 'cottages' in Ontario, Canada (1951). While in many ways this research set the pattern for future second home studies there was a gap of ten years or more before geographers returned to the subject. The greater importance of vacation homes in Scandinavia (Aldskogius, 1967; Norrbom, 1966), where a third or more of all households had access to second homes in the 1960s, meant that the phenomenon was better understood there than for example in Britain where perhaps only 2 per cent of the population owned second homes (Bielckus, Rogers and Wibberley, 1972; Dartington Amenity Research Trust, 1977).

Studies of second-home development in advanced societies have tended to take on a common format. At the national level the concern has been to measure concentrations of second homes, to trace the links between second-home locations and first-home origins and to develop a social and economic profile of the second-home user. Important examples of such research for particular countries, generally produced in the early 1970s, are those for France by Clout (1969, 1971) and for the United States by Clout (1972) and Ragatz (1970). A comprehensive review of second homes in several countries in Europe and North America, together with an assessment of some of the problems which they create, can be found in Coppock (1977). This pattern has also been followed by a host of regional and local studies of second homes. In addition, case studies of second-home development have particularly concentrated upon the problems caused by second homes, especially those relating to the social and economic contrasts between incomers and indigenous populations.

Second homes provide what is probably the clearest instance of the juxtaposition of the new and the old, of the urban and the rural, which is characteristic of those Western societies which have seen the demographic shifts of the latter half of the twentieth century. As such, the issues which they raise are particularly sensitive. In a country like Britain where planning controls upon construction in the countryside have limited the possibilities for new, purpose-built development, the pressures induced by the demand for second homes are directed to the existing housing stock. Accordingly the main concerns here are for the inequitable effects on housing access and the adverse consequences for poorer local people when faced with strong competition from outsiders (eg, Bennett, 1976). These problems can be exacerbated when cultural contrasts are also involved. In Wales in particular second homes have been a distinct focus for nationalist sympathies and for concern over the loss of regional identity and language (Bollom, 1978), leading to arson and subversive activity (Gilg, 1981).

In other countries, notably in North America and in Australasia, where land resources are less of a constraint and where control upon development is far less rigorous, the concerns are more for the environmental excesses of second-home growth. In particular the subdivision of land for second-home properties and thus its removal from full economic use in agriculture has been a constant issue. In Australasia the trend towards combining second homes with some form of part-time farming has given rise to so-called 'rural retreats' (Wagner, 1975). The growth in this form of second-home development has now become so large that there have arisen serious problems of water supply and contamination and of landscape destruction, with the result that controls are being introduced on future construction in some areas (Williams, 1976).

Before leaving this topic it should be noted that the recreational use of rural housing can have its positive as well as its negative aspects. Particularly in rural areas where there exists a surplus of rural housing, generally because of a long tradition of depopulation, there is the potential for significant economic development if surplus housing is used for recreational purposes. France provides a case where this situation has been turned to distinct advantage. Since 1954 the French government has attempted to stimulate rural development by encouraging rural people to improve surplus housing and rent it to tourists (Wrathall, 1980). Grants and low-interest loans are available to bring property up to the required standard. While each *department* controls and services the *gîtes* in its area, there is a national federation which maintains an overall view. By 1979 some 28,000 properties were in service as *gîtes* in over 4,000 villages of France, providing a significant injection of additional income to the rural economy and bolstering the existing rural population.

Rural Housing, Planning and Policy

Rural housing as a major component of 'micropolitan infrastructure' (Tweeten and Brinkman, 1976) inevitably plays a crucial role in policies and programmes for rural development. In general, policies for rural housing carried out by regional or national government agencies are designed to fulfill one or more of three broad objectives:

a) as an adjunct to agricultural policy and as a complement to price supports for agricultural commodities in improving agricultural production and so indirectly raising rural incomes;
b) as a means towards wider rural and regional development in broadening the rural economic base and improving the welfare status of poorer groups in the countryside;
c) as a vehicle for environmental protection and conservation.

The means by which such policies, whatever their objectives, may be carried out are correspondingly wide. Again, three main approaches are discernible: direct action by the state (usually a local government agency or special rural agency) to build housing; indirect action by the state in the way of incentives for rural construction and improvement and in particular by the provision of grants and credit; and thirdly the imposition of systems of control, usually through the agency of land planners, which are designed to safeguard and improve the rural environment.

The continued importance granted to the agricultural function of rural areas which has frequently been mentioned in this chapter has meant that, both in North America and in Western Europe, policies for rural housing have generally been formulated in line with the first of the objectives noted above. Implicit in many policies for agriculture is the idea that farming performance is directly related to the conditions of the living environment. Yet, although this relationship has been noted by several workers (e.g. Frawley (1973) in the context of Ireland), it has yet to be properly formalised in many agricultural policies. Fennell (1981) has criticised the Common Agricultural Policy of the EEC for concentrating on price support at the expense of improving living conditions and Pratschke (1981) has suggested that the European Social Fund could provide a suitable vehicle for programmes of rural housing improvement.

In the United States such programmes have taken a more central role in agricultural and rural development policies. In particular, the Farmers Home Administration (FmHA) has been responsible since 1949 for providing aid, mainly in the form of mortgage loans, for rural housing projects (Malotky, 1969; Nelson, 1975). While this aid was originally only available to farmers, it has over the years been extended to cover the needs of non-farming rural people (Bird and Kampe, 1977).

The FmHA sometimes acts alone and at other times joins with other federal and state agencies for rural housing development. An example of such collaboration is the demonstration project on rural housing in California which involved the FmHA, the federal Department of Housing and Urban Development (HUD) and the California Department of Housing and Community Development (Bradshaw and Blakely, 1979). However, the world recession in the late 1970s, coupled with a significant change of political emphasis with regard to intervention by government, seems likely to restrict the activities of agencies like the FmHA in the future.

Programmes for rural housing development, while generally the responsibility of government, are not always so. Unlike urban housing, however, the rural sector is rarely an attractive proposition for private commercial involvement and so private development designed to improve existing housing or to provide low-cost housing in rural areas is only usually possible when carried out by a non-profit organisation or where a development is subsidised by a government agency. An example of the former is the Rural Housing Organisation operating in the west of Ireland, particularly in the counties of Clare, Limerick and Tipperary (Rural Housing Organisation, 1976). The RHO is a non-profit company which was established in 1972 with a commitment to repopulate villages in Ireland through the provision of housing. By 1976, 374 houses were completed, under construction or planned.

In Britain some government agencies have performed a similar role in conjunction with the private sector. The Highlands and Islands Development Board in Scotland (Highlands and Islands Development Board, 1974) and more recently the Development Commission (Development Commission, 1981) in England are two examples of such agencies that have been active in rural housing construction as part of a much broader strategy of rural development.

The third objective, that of regarding rural housing in the context of land-use planning and environmental protection is one that has been particularly characteristic of Great Britain. The development of a strong land-use planning system after the Second World War coupled with a steadily decreasing agricultural population has combined to make the control of housing in the countryside at least as important as the provision of such housing. Through the medium of strategic planning policies (e.g. Woodruffe, 1976; Cloke, 1979) and the application of a system of development control which has particularly stressed the conservation of attractive rural housing, the quantity of construction in rural areas has been severely controlled. From the viewpoint of access to housing and the circumstances of the poorer elements of the rural community, this policy of constraint has in recent years been criticised on the grounds that it has caused the price of housing to rise beyond the means of many rural people (Dunn, Rawson and Rogers, 1981; Shucksmith, 1981).

Conclusion

It can be seen that rural housing in the advanced Western societies no longer exists in isolation from the rest of the economy. By virtue of increased mobility and the growing interaction between the residual elements of the agricultural sector and its urban-industrial counterpart, the rural housing stock has taken on a new set of functions and a new set of problems. The demographic changes of the latter half of the twentieth century which have led to the repopulation of rural areas have now added a social dimension to these functions and problems. The consequences of a further development in this direction if micro-technology continues to advance seem likely to increase the relative importance of rural housing by the end of the century.

References

Aldskogius, H. (1967), 'Vacation Home Settlement in the Siljan Region', *Geografiska Annaler*, 49 (2), 69-95

Beale, C.L. (1977), 'The Recent Shift of United States Population to Non-metropolitan Areas, 1970-75', *International Regional Science Review*, 2 (2), 113-22

Beale, C.L. (1980), 'The Changing Nature of Rural Employment', in Brown, D.L. and Wardell, J.M. (eds.), *New Directions in Urban-Rural Migration: the Population Turnaround in Rural America* (Academic Press, London)

Bennett, S. (1976), *Rural Housing in the Lake District* (University of Lancaster, Lancaster)

Bielckus, C.L., Rogers, A.W. and Wibberley, G.P. (1972), *Second Homes in England and Wales*, Studies in Rural Land Use no. 11 (Wye College, University of London, Ashford)

Birch, S.P. (1974), *Farmhousing in the United States Corn Belt in 1970: Trends and Problems Revealed by Housing Census Analysis* (Department of Geography, University of Southampton)

Bird, R. and Kampe, R. (1977), *25 years of Housing Progress in Rural America*, Agricultural Economic Report no. 373 (US Department of Agriculture, Washington, DC)

Bollom, C. (1978), *Attitudes and Second Homes in Rural Wales*, Social Science Monograph no. 3 (University of Wales Press)

Bradshaw, T.K. and Blakely, E.J. (1979), *Rural Communities in Advanced Industrial Society: Development and Developers* (Praeger)

Brown, D.L. and Wardell, J.M. (1980), *New Directions in Urban-rural Migration: the Population Turnaround in Rural America* (Academic Press, London)

Brunhes, J. (1920), 'Types regionaux de maisons' in G. Hanotaux (ed.), *La geographie humaine de la France* (Paris)

Clark, D. (1980), *Rural Housing in East Hampshire* (National Council of Social Service, London)

Cloke, P. (1979), *Key Settlements in Rural Areas* (Methuen, London)

Clout, H. (1969), 'Second Homes in France', *Journal of the Town Planning Institute*, 55, 440-3

Clout, H. (1971), 'Second Homes in the Auvergne', *Geographical Review*, 61, 530-3

Clout, H. (1972), 'Second Homes in the United States', *Tijdschrift voor Economische en Sociale Geografie*, 63, 393-401

Clout, H. (1974), 'The Growth of Second-home Ownership: an Example of Seasonal Suburbanisation' in Johnson, J.H. (ed.), *Suburban Growth: Geographical Processes at the Edge of the Western City* (Wiley, London)

Coppock, J.T. (ed.) (1977), *Second Homes: Curse or Blessing?* (Pergamon Press, Oxford)

Dartington Amenity Research Trust (1977), *Second Homes in Scotland* (Dartington Amenity Research Trust, Totnes)

Deaton, B.J. and Hanrahan, C.E. (1973), 'Rural Housing Needs and Barriers: the Case of Central Appalachia', *Southern Journal of Agricultural Economics*, 5 (1), 59–67

Demangeon, A. (1920), 'L'habitation rurale en France: essai de classification', *Annales de Geographie*

Department of the Environment (1978), *English House Condition Survey, 1976* (HMSO, London)

Development Commission (1981), *Thirty-ninth Report* (HMSO, London)

Dunn, M., Rawson, M. and Rogers, A. (1981), *Rural Housing: Competition and Choice* (George Allen and Unwin, London)

Fennell, R. (1981), 'Living Conditions in Rural Areas in the European Community', *Paper given to the Third Congress of the European Association of Agricultural Economics* (Belgrade)

Frawley, J. (1973), 'Social Factors Related to Farming Performance', *Paper read to the Agricultural Economics Society of Ireland*

Gasson, R. (1975), *Provision of Tied Cottages*, Occasional Paper no. 4 (Department of Land Economy, University of Cambridge, Cambridge)

Gilg, A. (1981), *Countryside Planning Yearbook, 1981* (Geobooks, Norwich)

Green, R.J. (1971), *Country Planning: the Future of the Rural Regions* (University of Manchester Press, Manchester)

Hereford and Worcester County Council (1978), *Rural Community Development Projects* (Hereford and Worcester County Council, Worcester)

Highlands and Islands Development Board (1974), *Rural Housing in the Highlands and Islands*, Occasional Bulletin no. 5 (Highlands and Islands Development Board, Inverness)

Hoskins, W.G. (1970), *History from the Farm* (Faber and Faber, London)

Houston, J.M. (1964), *The Western Mediterranean World* (Longman, London)

Irving, B.L. and Hilgendorf, E.H. (1975), *Tied Cottages in British Agriculture* (Tavistock Institute for Human Relations, London)

Limmer, R. and McGough, D. (1979), *How Well are we Housed?: 5, Rural* (US Department of Housing and Urban Development, HUD-PDR-500, Washington, DC)

Mackenzie, A.M. and Martin, P.C. (1975), 'Farm Dwellings', *Scottish Agricultural Economics*, 25, 370–2

Malotky, L. (1969), 'Housing in the Rural Areas of the United States of America' in *Rural Housing: a Review of World Conditions, United Nations* (UN)

Marantz, J.K., Case, K.E. and Leonard, H.B. (1976), *Discrimination in Rural Housing* (Lexington Books, Farnborough)

Mikesell, J.J. (1973), 'Mobile Homes: an Important New Element in Rural Housing', *Agricultural Finance Review*, 34, 18–23

Nelson, B.F. (1975), 'Rural Housing: a New Possibility for Local Housing Authorities', *Journal of Housing*, 32 (5), 227–32

Norrbom, C-E. (1966), 'Outdoor Recreation in Sweden', *Sociologia Ruralis*, 6 (1), 56–73

Office of Population Censuses and Surveys (1981), *Census 1981: Preliminary Report, England and Wales* (HMSO, London)

Ogden, P.E. (1978), 'Analyse multivariee et structure regionale: transformations socio-economiques recentes dans le Massif Central de l'est', *Mediteranée*, 3, 45–58

Pitie, J. (1969), 'Pour une geographie de l'inconfort des maisons rurales', *Norois*, 63, 461–90

Pratschke, J.L. (1981), 'Rural and Farm Dwellings in the European Community',

Irish Journal of Agricultural Economics and Rural Sociology
Ragatz, R.L. (1970), 'Vacation Housing: a Missing Component in Urban and
Regional Theory', *Land Economics*, 46, 118–26
Rogers, A.W. (1971), 'Rural Housing' in Cherry, G.E. (ed.), *Rural Planning
Problems* (Leonard Hill, London)
Rural Housing Organisation (1976), *A Case Study of Village Development in
Ireland* (Shannon Free Airport Development Company Ltd, Shannon)
Shucksmith, M. (1981), *No Homes for Locals?* (Gower, Farnborough)
Trewartha, G.T. (1948), 'Some Regional Characteristics of American Farmsteads',
Annals of the Association of American Geographers, 38
Tweeten, L. and Brinkman, G. (1976), *Micropolitan Development* (Iowa State
University Press, Ames)
USDA (United States Department of Agriculture) (1971), *The Economic and
Social Condition of Rural America in the 1970s* (Economic Research Service,
USDA, Washington)
Wagner, C. (1975), *Rural Retreat: Urban Investment in Rural Land for
Residential Purposes* (Australian Government Publishing Service, Canberra)
Whitby, M. and Hubbard, L. (1982), 'The Urban-rural Income Gradient and the
Pressure of Demand for Labour', *Regional Studies*, forthcoming
Williams, M. (1976), 'Rural Planning is not so Simple', *Royal Australian Planning
Institute Journal*, July/October, 22–6
Wolfe, R. (1951), 'Summer Cottages in Ontario', *Economic Geography*, 27,
10–32
Woodruffe, B.J. (1976), *Rural Settlement Policies and Plans* (Oxford University
Press, Oxford)
Wrathall, J.E. (1980), 'Farm-based Holidays', *Town and Country Planning*, 49
(6), 194–5

6 TRANSPORT AND ACCESSIBILITY

D.J. Banister

Introduction

Concern over the levels of accessibility in rural areas has increased during the last twenty years. The rise in car ownership has been reflected by a decline in the demand for local services including public transport. New responsibilities have been placed on local authorities to mitigate the worst effects of these changes, but their actions have been limited particularly in the light of the general cutbacks in resource availability.

This chapter first defines transport mobility and accessibility, making reference to the studies which have proliferated on these issues over the last twenty years. Next the recent changes which have taken place in rural areas are outlined, with particular emphasis on the transport issues. The conclusion here is that the context is dynamic and that the nature of rural problems has changed rapidly and is likely to continue to do so. These changes are then related to the political response as two recent Transport Acts have marked the most significant switches in policy with respect to rural transport for the last fifty years. The alternatives available to the planners and transport operators are stated, and the conclusions that are drawn seem to indicate that the role for conventional public transport in rural areas may be limited. Policy should explicitly consider all alternatives, including better use of the private car and non-transport options. The chapter concludes with a set of proposals for transport and accessibility in rural areas.

Transport Mobility and Accessibility

Any discussion of transport in rural areas revolves around the twin concepts of mobility and accessibility. Mobility can simply be defined as the ability of an individual to move about. This ability can operationally be examined in two parts, the first of which relates to the amount of travel which is actually made in terms of the numbers of trips by all modes (including the walk mode) for all purposes (including multipurpose trips). The second part covers the ease of movement, which is much harder to quantify. It would include personal characteristics (age, sex, socio-economic variables, physical disability), whether a car is available or not (either as a driver or a passenger), the ability to drive a car and the availability of public transport services and other modes (motorcycle and pedal cycle).

Personal mobility has, then, been defined as a function of two variables. The first is the tangible realisation of travel demand in terms of actual movement, whilst the second is a personal categorisation which acts as a constraint on the first, and can be interpreted as the potential for movement. With this definition, it is recognised that not everybody wants or requires the same level of mobility. This point is apposite at the time when the government has directed the counties to prepare annual public transport plans which include policies to 'promote the provision of a co-ordinated and efficient system of public transport to meet the county's needs' (G.B., Department of Transport, 1978a).

Increasingly however, the focus of attention has switched from the problems of mobility to those of accessibility. Transport is now seen as the means to an end and not an end in itself. People travel because they derive benefits at their destination which exceed the costs of travel; thus if transport is treated as a derived demand, consideration must be given to the distribution of facilities as well as the transport links and the spatial and social characteristics of the population. Transport accessibility can be defined as the ability of people to get to or be reached by the opportunities which are perceived to be relevant to them (Jones, 1981).

Conventionally, accessibility can be considered in two ways. Locational accessibility is either a comparative measure which weighs units of separation against the number of destinations, or a composite measure which combines the two factors into a single index. Comparative measures of accessibility are numerous and indicate the increasing number of opportunities which become accessible with distance (Briggs and Jones, 1973; Mitchell and Town, 1977; Joint Working Group of the TRRL, 1981). Composite accessibility is usually based on the gravity model where accessibility is a function of the attractiveness of the destination and the distance to be travelled (Hansen, 1959; Baxter and Lenzi, 1975). The applications are common in transport planning, where the variations on the basic theme are numerous (Johnson, 1966; Daly, 1975; Pirie, 1979; Morris *et al.*, 1979).

Secondly, there is personal accessibility where the unit of study is the individual rather than the location. Here the individual is limited by the available modes of travel and communication, and these are modified by the person's perception and determined by the individual's activity demand pattern (Wachs and Kumagai, 1973). Perhaps the best example is the time-space school of geography at Lund, where Hägerstrand (1970, 1972) and his team have concentrated on the opportunities which are available rather than those which are actually used (Lenntorp, 1981). Each individual has his own 'action space' which limits the activities he can take part in, and the objective of this approach is firstly to determine the dimensions of the 'action space' and secondly to expand it or to place more opportunities within it.

Accessibility is very unequally distributed in rural areas, both between different locations and within individual locations. Research has almost exclusively concentrated on the demand and 'need' elements of particular consumers. The producers or suppliers of services have not yet come under the researcher's microscope (Moseley, 1981). Social group studies have proliferated; these include the elderly (Hopkin *et al.*, 1978), housewives (Pickup, 1981), non-car-owning households (Koutsopoulos and Schmidt, 1976) and low-income groups (Davis and Albaum, 1972). Perhaps the best-known empirical studies of the mobility and accessibility problems faced by particular groups have been those of Hillman and his colleagues at Political and Economic Planning (Hillman *et al.*, 1973, 1976). Area based studies have also been popular (Moseley *et al.*, 1977; Banister, 1980a), and particular modes, such as community transport (G.B., Department of Transport, 1978b) or unconventional public transport (Balcombe, 1980) have provided another focus for analysis. Finally, some studies have attempted to set this particular concern within the context of other rural problems (Banister, 1980b; Shaw, 1979). In each case the research has concentrated on the problems of a particular individual or group within a particular area with the twin purpose of identification of the disadvantaged and the determination of the nature of the problem (Town, 1980). There has been relatively little attention paid to the evaluation of alternative policies that could be used to alleviate the problems identified. These issues are returned to later after the dynamics of rural areas have been introduced together with the responses from the transport policy-makers.

The Dynamics of Change

Significant changes have taken place in rural areas with respect to the population, the distribution of facilities and the transport links. It is difficult to generalise, but an attempt will be made to highlight certain key issues which have a special relevance to transport provision and accessibility. The first dimension of change is the social one. From the 1981 census, there is some evidence of population growth in urban fringe areas, particularly in the southern part of Britain (see Chapter 4). Additionally, many upland areas have reversed the population declines of the 1960s (Randolph and Robert, 1981). However, the overall changes conceal the selective nature of the movers out of rural areas and the movers back into rural areas. The net result is that one is not left with a balanced profile of the total population but people with different life-styles and aspirations. There is the old established resident who may be elderly (about 25 per cent of the population in the remote rural areas are elderly, as compared with a national average of 17 per cent) and who has had to accommodate to declining levels of service; on the other hand there is the recent migrant who may have urban-based life-styles and only resides in the rural area but plays no active part in the rural community. In effect, one has communities within communities (Pahl, 1964).

The one persistent feature with respect to the provision of facilities has been the trend towards concentration into fewer, larger units. These units, particularly shops, are usually located in the urban centres and not locally, with the result that the distances to be travelled will be increased. There is also some evidence of school closure in rural areas, particularly in the primary school sector where a minimum of 50-60 pupils was considered by the Plowden Report (G.B., Department of Education, 1967) to be a 'viable' unit. Health-care facilities have become characterised by group practices with large catchment areas and hospital facilities have become both more centralised and specialised (Rigby, 1977). However, there is evidence to suggest that a reversal is now taking place with the growth of community hospital facilities. Many of these changes have taken place without a thorough considera-tion of the additional transport costs involved. Concentration has also resulted in the closure of local facilities, including shops, post offices and petrol stations (Association of County Councils, 1979; Taylor and Emerson, 1981). Petrol prices are typically about 15 per cent higher in rural areas and oil companies have been pursuing a policy of closure for small rural petrol stations; Shell and Esso closed 2,000 rural petrol stations in 1980 (Motor Agents Association, 1981). All these changes have resulted in a decline in the levels of local accessibility with an increase in the requirements for transport.

Table 6.1: Journey Lengths in Rural Areas — By Stage in Kilometres

Purpose	Rural[a] areas	Mean for all areas	% difference
Work	7.91	5.75	+ 36
Education	4.10	2.56	+ 60
Shopping	6.24	3.44	+ 81
Social	9.95	7.40	+ 34
Escort	7.79	4.88	+ 60
Other	15.60	10.94	+ 43
All purposes	8.85	6.30	+ 41

Journey Lengths in Rural Areas — By Trip Rates and Mean Trip Length in Kilometres

Location	Trip rates per person per week	Mean trip length (km)	Total distance travel per week (km)
Rural areas[a]	16.85	10.81	182.23
All areas	17.92	8.24	147.74

Note: a. Rural areas defined as those areas with a population under 3,000.
Source: Unpublished National Trust Survey Tabulations, 1975/6.

Car ownership levels have continued to rise despite increases in the prices of fuel and new vehicles, with the only change being a reduced rate of increase. In rural areas, about 80 per cent of households own at least one car, and the total level of vehicle ownership is much higher if bicycles and motorcycles are included: in south Oxfordshire the level was 1.91 vehicles per household in 1977 (Banister, 1980b). Car ownership levels tend to correlate positively with household income and negatively with population density (Rhys and Buxton, 1974). Thus a household with a given income is likely to have a higher propensity to own a car in a rural area than in an urban area.

As mentioned earlier, journey lengths in rural areas are greater than those in all areas, with journey stages being about 40 per cent higher overall. Shopping stands out as the trip purpose with the greatest difference (Table 6.1), but in all cases rural people have to travel further. So that although trip rates are slightly lower, the greater mean trip lengths give rural residents a total distance travelled per week nearly 25 per cent above the mean for all areas. The time taken to make a journey may not differ by so much, as journey speeds are typically higher in rural areas.

Public transport patronage has declined from peak levels in 1952 by about 50 per cent. At that time nearly 40 per cent of all vehicular

Table 6.2: Modal Split in Rural Areas

Mode	Distance travelled — percentage	
	National Travel Survey, 1965	National Travel Survey, 1975/6
Car or van as driver	39	48
Car or van as passenger	32	33
Motorcycle and bicycle	3	2
Walk (over 1.6km)	2	1
Local bus	13	5
Long-distance bus or coach, works or school bus, train, other	11	10

Source: GB, Department of Transport, 1978c.

passenger kilometres were by bus, but now the level is about 8 per cent in rural areas. There are several reasons, the principal one being the increase in car ownership. Oldfield (1979) has estimated that for every additional car, 300 bus trips were lost each year and that the increase in car ownership accounted directly for about 45 per cent of observed public transport patronage decline. The consequence of the decline in demand has been that bus operators have reduced the level of service provided particularly frequency, and during the last four years alone fares have been raised by about 30 per cent in real terms. However, the fall in the annual vehicle kilometres has not been of a similar order (a 19 per cent decline 1969-80). Buses are now running emptier and so vehicle productivity has fallen despite increases in staff productivity. Overall, the number of passenger journeys on services run by the National Bus Company have declined by 38 per cent (1969-80).

Rural railways account for about 1,000m passenger kilometres a year with gross revenue amounting to £26m a year and direct expenses totalling £44m a year (Hodge, 1979). The rural railway network was halved under the Beeching Plan (British Railways Board, 1963) to about 5,000 kilometres, and their present contribution to total rail travel is about 4 per cent. It seems that railways are likely to continue to play a minor role in rural areas despite innovations such as the British Rail/Leyland lightweight experimental vehicle.

A quiet revolution seems to be taking place in the countryside and some of the principal issues have been highlighted. Accessibility for those who have acquired or already own a car has been increased or maintained, but for those with no access the situation is the reverse. Local facilities have closed and distances to be travelled have increased, but the vital transport links have also been curtailed. Thus, as in other sectors, there seems to be a move from the public provision of transport (eg, the bus) towards the private provision of transport (eg, the car or the bicycle). Table 6.2 illustrates these conclusions, in terms of kilometres travelled by mode.

Policy Responses

To understand the present situation, it is essential to mention the changes that have taken place in transport policy and how this has and is likely to affect the provision of transport services in rural areas. As mentioned in the previous section, public transport patronage levels have declined over the last thirty years. The problem was first identified in the Jack Committee's Report (Ministry of Transport, 1961) on rural transport services where it was concluded that 'the present and probable future levels of rural bus services are not adequate to avoid a degree of hardship and inconvenience sufficient to call for special steps'. The social function of transport services was recognised, and the suggestion was made for selective direct financial assistance. The Jack Committee noted that there was nothing inherently wrong with the public transport operations, the main reason for decline was the growth in car ownership. Subsequent evidence (Oldfield, 1979) supports this claim.

The 1968 Transport Act, although concentrating mainly on urban transport issues, set up a system of direct grants for unremunerative services with central government paying half (Section 34) provided that the service covered at least 50 per cent of its costs. The Act also amalgamated the nationally-owned Transport Holding Company and the British Electric Traction Company into the National Bus Company, which now operates about one in three of buses in England and Wales, including most stage carriage services in rural areas. There was some relaxation of control over bus services as school contract services were now permitted to carry fare-paying passengers on certain routes provided that there was excess capacity (Section 30). Other measures included the increase in fuel-tax rebates, grants for new buses, and the introduction of concessionary fares.

Since then, the situation for the public transport operator has further deteriorated, despite many studies of the 'rural transport problem' initiated either by central government (eg, Department of the Environment, 1971a, b) or by other agencies (eg, Clout *et al.*,

1972). The findings from these studies emphasised the difficulties for operators. Demands for public transport services were so disparate and varied that it was almost impossible to match them together to form any sensible public transport load.

Further measures were taken in the Local Government Act (1972), when the non-metropolitan county councils were directed to develop policies which would promote 'the provision of a co-ordinated and efficient system of public passenger transport to meet the county's needs' (Section 203). The local authorities were now to administer the distribution of subsidies and in many areas the 'needs' of the local communities were reviewed. Each county was also directed to submit an annual Transport Policy and Programme (TPP) which would contain an estimate of transport expenditure for the following financial year, a statement of transport objectives, a strategy in outline for the next five years, a statement of short-term investment proposals and a summary of the past expenditure. The TPP forms the basis for the allocation of the Transport Supplementary Grant (TSG), and took effect from the financial year 1975/6.

The TSG for the whole country amounts to £344m (at November 1979 prices) for 1981/2 with nearly 65 per cent going to the metropolitan counties. Since its introduction, both the levels of accepted expenditure and the actual TSG have fallen in real terms by 30 per cent and 50 per cent respectively. In the last year there has been a switch back to capital expenditure (35 per cent of accepted expenditure) at the expense of road maintenance and public transport revenue support (48 per cent and 18 per cent of accepted expenditure respectively) (G.B., Department of Transport, 1981).

The TSG is channelled to the operators through the county councils, whilst the fuel-tax rebates and new bus grants are paid direct to the operator.

This emphasis on the metropolitan areas was revised in the 1978 Transport Act with a modest shift in resources towards rural services. The social importance of rural transport services was recognised with an undertaking to support services and an invitation from the government for the shire counties to revise their revenue bids upwards. About a half responded and almost all the revised bids were accepted. Each county was also required to submit an annual County Public Passenger Transport Plan (PTP) which would provide a clear statement of objectives and policies for public transport together with the financial and resource commitments. Public transport in rural areas had taken on an explicit social function in that services should be provided to meet the 'needs' of the residents. The government gave no guidelines as to how these 'needs' should be measured, and most counties opted for some minimum levels of service, often related to population size. The standards range from a relatively high level with a complex hierarchy of services (eg, Bedfordshire) to simple service thresholds (eg, East Sussex). At the other end of the spectrum some local authorities (eg, Oxfordshire) have made no real mention of 'needs' in their PTP as local people are expected to identify their own particular transport problems and priorities.

A series of other measures was also introduced. These included a commitment by government to subsidise at least one half of the cost of all off-peak travel by the elderly, blind and disabled (G.B., Department of Transport, 1979), the exemption of minibuses of between eight and sixteen seats from public service licences provided that they were used by voluntary organisations, and the legalisation of regular car-sharing schemes for payment, provided that a 'profit' was not made.

During this period two other important activities were taking place. First the government established sixteen rural transport experiments in Devon, North Yorkshire, South Ayrshire and Dinefwr (in Dyfed). The areas selected were all 'deep' rural areas where conventional services were never likely to be viable, and the purpose of the experiments was to test the impact of a relaxation of the licensing laws so that service innovation could take place. The experiments included flexible route services, community services, shared hire cars, organised transport services by private car, hospital services, post-bus services and an emergency car service (Balcombe, 1980). None of the services operates without a subsidy, but at the lowest levels of demand, voluntary car services seem to provide the best value for money. The reason for this is that there are no direct wage costs to pay. Where demand is higher the post-bus, the community bus, and the shared hire car services can provide a public transport service at roughly equivalent levels of subsidy. All experiments are now completed but ten services are still in operation with financial support from local authorities (Coe and Fairhead, 1980).

The second change was the Market Analysis Project (MAP), introduced by the National Bus Company as a selective service planning tool. It arose from the necessity for rationalisation and the avoidance of a piecemeal approach to service provision which was typical of many operations. MAP looks at the entire network in and around a town or an area within a county with a particular view to the reduction in the peak vehicle requirement. The effect has been to match service provision more closely with expressed demand, with changes being introduced in routes and frequencies so that say 90 per cent of demand could be met by either substituting one route for two or through rescheduling of services. All National Bus Company subsidiaries have now completed their market analysis exercises (G.B., Department of Transport, 1978d).

Most recently the Transport Act 1980 has introduced radical changes in the provision of public transport services, particularly in rural areas with service deregulation. Small vehicles (fewer than 8 seats) no longer require any licences. Long distance bus services (with a minimum journey length over 48 kilometres) do not now require a road service licence. This change has led to the formation of British Coachways, a consortium of private operators, who are now in direct competition with the National Bus Company. Some fares have been halved and service frequencies increased on the intercity bus routes; there have been indirect benefits to rural communities located on or near these routes.

For other services, road service licences are now granted provided that the new service is not against the public interest. Operators also have more freedom to charge the fares that the market can bear. County councils have the opportunity to apply for trial area status in part or all of their county where there would be no road service licences, and operators can run services on any routes they desire. So far trial area status has been granted to the old county of Hereford-shire and to north-west Norfolk. For rural areas, these policies may result in a significant reduction in public transport services. Many routes were dependent on cross-subsidisation from profitable services in town or between towns, but now each route is treated on its own merits with intense competition on the profitable routes. The explicit social function of rural public transport services highlighted in the 1978 Transport Act has now reverted to economic criteria with the intro-duction of the 1980 Transport Act (Table 6.3).

The feeling within the bus industry is that in the near future many rural areas may have a very limited public transport service (National Bus Company, 1981). There have been unprecedented changes in policy towards the provision of public transport in rural areas from a situation of tight control to one of deregulation. In particular the role of the traffic commissioners, almost unchanged since the 1930s, has now become one of facilitating new services rather than maintaining existing services.

Table 6.3: Principal Transport Policy Changes and their Impact on Rural Areas

	Date	Recommendations and actions
Jack Committee's Report on Rural Bus Services	1961	Selective direct financial assistance to unremunerative services.
		Fuel-tax rebate.
Beeching Report on the Reshaping of British Railways	1963	Closure of 5,000km of rural railways.
Series of surveys in six areas	1963	To examine the problems of and the demand for rural transport services.
Transport Act	1968	System of direct grants with central government paying half where the services covered half its operating costs.
		Revenue grants to cover losses on unremunerative rail services.
		National Bus Company set up.
		School service contracts could now take fare-paying passengers if there is excess capacity.
		Fuel-tax rebates increased.
		New bus grants introduced.
		Concessionary fares.
Local Government Act	1972	Co-ordinating function for the county councils through the Transport Policies and Programmes.
		County councils to administer the distribution of the Transport Supplementary Grant.
Passenger Vehicle (Experimental Areas) Act	1977	Relaxation of the licensing laws in certain areas so that innovative services could be introduced.
	1977	Rural Transport Experiments set up in four 'deep' rural areas.
Transport Act	1978	County Public Transport Plans to co-ordinate passenger transport to meet the 'needs' of the public.
		Guidelines set for concessionary fares.
		Minibuses (8–16 seats) exempt from Public Service Vehicle Licences and Road Service Licences, provided that the drivers were from approved voluntary organisations.
		Traffic commissioners permitted to introduce short-term Road Service Licences.
		Car sharing allowed for payment.
Transport Act	1980	Major changes in bus licensing — deregulation.
		Small vehicles (fewer than 8 seats) no longer classified as Public Service Vehicles.
		Express services (minimum journey length over 30 miles) no longer required Road Service Licences.
		Road Service Licences 'create a presumption in favour of the applicant'.
		Trial Areas can be designated where there are no Road Service Licences.

The Choices Available

From this review of the context and policy changes over the last thirty years, it seems that the problems of accessibility have increased rather than decreased. Distances that have to be travelled to reach activities have lengthened with the closure of local facilities and the concentration of other facilities in the larger centres. Public transport services have been reduced and buses are now running emptier. The response of many people has been to purchase a car or second car or some other means of private transport; this seems to be a prerequisite for anyone moving into rural areas. The problems of accessibility are minor if the individual has access to his own means of transport at any time as their main constraint is one of travel time. However, it is those individuals without the ready availability of transport who are inaccessible, even though there is evidence of considerable flexibility in trip making to ensure the maximum use of available cars with lift-giving and informal car sharing (Banister, 1980b). Nevertheless their problems have been multiplied with the decline in local services and public transport. As Hillman *et al.* (1973) have indicated, this section of the population is likely to remain substantial as it includes those who are too old or too young to drive, those who have some disability, those who are in the low-income groups and those who are in households with one car where the head of the household takes the car to work leaving the other members of the family without its use during the day. The population may become polarised into those who do and those who do not have ready access to a car (Table 6.4).

Policy has responded in a variety of ways; the public transport operator now has greater freedom in service provision, some of the restrictions on the use of the private car for lift-giving have been removed, and there is more scope for the introduction of experimental services (eg, community buses, post-buses). Nevertheless, in the short term there is likely to be a decline in the resources available for public transport support both from central government and the county councils. One alternative would be to raise the support locally through a precept on the rates or some form of local taxation, but this is unlikely in the Conservative-controlled shire counties. Another possibility would be to increase the borrowing requirement of the National Bus Company, but again this is unlikely. In fact the government is continuing to reduce the levels of Transport Supplementary Grant in real terms, and new bus grants (worth £66m to bus operators in 1979/80) are being phased out by 1983.

Table 6.4: Car Availability in South Oxfordshire, 1977

	Percentage
Car available at any time	62
Car never available either for trips as a driver or a passenger	18
Car available at restricted times (eg, in evenings or at weekends)	20

Source: Banister, 1980b.

And so the operator has the dilemma of trying to maintain services on a reduced level of revenue support. As has been mentioned, the possibility of cross-subsidisation from urban services to rural ones has been eliminated by the 1980 Transport Act which effectively allows every route to be competed for on an individual basis. There seem to be five main options. The first would be to increase the levels of subsidy from the county council, but this is not feasible given current government policy on public expenditure cutbacks. The second would be to raise fares. These were kept at an artificially low level up to the mid-1970s, but since then they have been raised at a level above the rise in all levels of consumer expenditure. However, it now seems that elasticities of demand with respect to fares are increasing towards unity. The extra revenue gained from increasing fares is outweighed by losses in patronage (Bly, 1976). Reductions in levels of service could result in cost savings with the withdrawal of peak service vehicles (Tyson, 1972), but in rural areas with existing frequencies already at very low levels this alternative may result in no service. Route rationalisation has usually been carried out with the previous option. The Market Analysis Project (MAP) of the National Bus Company has already been mentioned and it is now complete for England and Wales. Finally, operations could be handed over to private bus and coach companies. Private operators have mean fare levels that are 25 per cent lower than those of the public operators, and given comparable levels of revenue support the differences might be 30-40 per cent. The differences are due to lower unit costs rather than higher loadings, and to no commitment to supply extensive stage networks (Tunbridge and Jackson, 1980).

The consumer or rural resident is the last link in the chain. No one has yet invented a means of transport that is of a higher quality than the private car, at least to the user. Its convenience and flexibility are

Table 6.5: Travel Patterns in South Oxfordshire, 1977

Trip rates Trip/person/day	10-24	Age group 25-59	60 +	Total	Licence- holders	Non- licence- holders
Mean[a] (1)	3.10	2.78	1.65	2.60		
(2)	3.58	3.39	2.85	3.33		
By mode (%)						
Walking	23.2	13.0	30.1	16.9	9.1	38.4
Car driver	20.5	59.5	44.7	50.7	70.0	0.0
Car passenger	28.2	18.2	18.1	19.9	13.7	36.8
Public transport	13.7	5.2	3.6	6.4	2.1	14.6
Other[b]	14.4	4.1	3.5	6.1	5.1	10.2

Notes: a. (1) mean for all members in sample; (2) mean for all members who actually made trips during the survey period. b. Includes pedal-cycle, motor-cycle, goods vehicle, other.

Source: Banister, 1980b.

unique. As such the car does not compete for patronage with public transport in rural areas, it serves a complementary function. No one with a car available would consider the bus as an alternative. The car is a superior mode on all counts, including cost (Webster, 1977), particularly if penalties are imposed on public transport users for the extra walk, wait and in-vehicle time required for travel. Public transport only serves those without access to a car. In Herefordshire, it has been estimated (White, 1978) that only 36 per cent of bus passengers come from non-car-owning households, and the remainder from car-owning households (principally those with no car available). The main users are the young, not the elderly, with the two main trip purposes being shopping (39 per cent) and education (26 per cent). The elderly and the low-income households travel less and tend to make a higher proportion of their trips on foot (Table 6.5).

One response would be to make better use of the available car. Another would be to purchase a car or a second (or third) car. Ownership levels are already much higher than the national average, but with lower mean income levels (Thomas and Winyard, 1979), longer distances to travel and higher petrol prices, low-income households may find it difficult to make ends meet. There may be 'forced' ownership in that a car is a prerequisite to living in rural areas; similarly, once a car is acquired, the families may be reluctant to give it up. The final response could be to move out of the rural area into a neighbouring urban centre where there may be a wider range of jobs, better local facilities and higher quality public transport.

A Policy for Transport in Rural Areas

It seems opportune for a fundamental re-examination of the role of public transport in rural areas. People living in these areas have over the last twenty years benefited from a public transport service that is better than commercial criteria could support. Service retention has been justified mainly on social criteria. This position has changed with the 1980 Transport Act with each route now being judged individually. When placed in the context of a decline in public expenditure, particularly on revenue support, an increase in unemployment and short-time working, high inflation and energy costs, the future seems bleak.

In planning for rural areas it is important to take a broad perspective that includes all sectors to determine whether the problem of accessibility is primarily a transport one. Decisions affect both the 'need' to travel (eg, the closure of a local shop) and the ability to travel (eg, the withdrawal of a bus service). Consultation and co-ordination between local authority departments may be limited and this may affect the availability of transport (eg, school buses for general use). Accessibility-related decisions may be taken with other objectives in mind. The post office, for example, is required to run a mail service, not a passenger transport service. Indirect effects, such as a school closure, may increase the demand for transport to school. Research has examined the problems of accessibility at the local level with specific solutions in specific situations. The conventional approach to the accessibility problem is to reconcile three conflicting objectives: low cost, good accessibility and a wide geographical coverage (Moseley, 1979). More important however may be the political and economic context within which decisions are made, particularly the effects that transport has on the well-being of the local economy and how these can be evaluated.

Radical alternatives to the current policy of incremental change should be examined. The options available for transport should not only include conventional public transport but the whole spectrum of alternatives as well as making better use of the private car. The role of the conventional bus may be restricted to the main inter-urban routes, with some rural communities benefiting from being on or near these routes. Secondly there is likely to be more extensive use of the private car, together with further increases in car ownership. The car may take on an explicitly social function with the introduction of shared car ownership and the village car. Pedal cycle and motorcycle use may also increase with access to certain minor roads restricted to pedal cyclists and local access. The public sector contribution may be in the form of limited demand-responsive services (eg, Dial-a-Bus) supplied at a premium price. These services would act as collectors to link the remoter areas to a town or bus or rail route. Perhaps there would also be some limited increases in dual purpose vehicles (eg, post-buses, delivery vehicles and mobile services). The major growth sector is likely to be the 'informal' services, such as car-sharing schemes, voluntary schemes (eg, good-neighbour services) and private car hire services that are tailored to individual requirements with no fixed scale of charges. It seems that the levels of demand for publicly provided transport services in rural areas are so dispersed and small in number that responsiveness to local requirements is the key (Banister, 1980a). The public sector may only have a limited role to play in satisfying the residual demands that cannot be met by private transport.

Finally there are the non-transport options for easing the problems of rural accessibility which can either act as alternatives or as complements (Moseley, 1979). Some have been mentioned, such as services to the people. Included here would be shopping deliveries as well as social or welfare visits. A location policy that favoured concentration rather than dispersal may reduce the number and length of trips required, while substitutes such as televisions, telephones, telecommunications, freezers and automatic dispensing may all reduce the demand for movement (National Council for Voluntary Organisations, 1980). Evidence on these alternatives indicates that the demand for travel may actually be increased rather than reduced (Clark, 1981). There is still an urgent requirement for a comprehensive rural accessibility policy that seeks to make the best use of all forms of communication, both physical and non-physical.

References

Association of County Councils (1979), *Rural Deprivation* (Association of County Councils Report, September)

Balcombe, R.J. (1980), *Summary and Conclusions. The Rural Transport Experiments*, Proceedings of a symposium held at the Transport and Road Research Laboratory, Crowthorne, November 1979, SR 584, 94–103

Banister, D.J. (1980a), 'Transport Mobility in Interurban Areas: a Case Study Approach in South Oxfordshire', *Regional Studies*, 14, 285–96

Banister, D.J. (1980b), *Transport Mobility and Deprivation in Inter Urban Areas* (Farnborough, Saxon House)

Baxter, R.S. and Lenzi, G. (1975), 'The Measurement of Relative Accessibility', *Regional Studies*, 9 (1), 15–26

Bly, P.H. (1976), 'The Effect of Fares on Bus Patronage', *Transport and Road Research Laboratory*, LR 733

Briggs, D.A. and Jones, P.M. (1973), 'Problems in Transportation Planning in the Conurbations: the Role of Accessibility', Paper presented to the Institute of British Geographers

British Railways Board (1963), *The Reshaping of British Railways* (London, BRB)

Clark, D. (1981), 'Telecommunications and Rural Accessibility: Perspectives on the 1980's' in Banister, D.J. and Hall, P.G. (eds.), *Transport and Public Policy Planning* (London, Mansell), 134-47

Clout, H.D., Hollis, G.E. and Munton, R.J.C. (1972), 'A Study of the Provision of Public Transport in North Norfolk', Occasional Paper No. 18 (Department of Geography, University College, London)

Coe, G.A. and Fairhead, R.D. (1980), 'Rural Travel and the Market for Public Transport in RUTEX Areas. The Rural Transport Experiments', Proceedings of a symposium held at the Transport and Road Research Laboratory, November 1979, SR 584, 3-18

Daly, A. (1975), 'Measuring Accessibility in a Rural Context', Proceedings of a seminar on Rural Public Transport, Polytechnic of Central London

Davis, C.S. and Albaum, M. (1972), 'Mobility Problems of the Poor in Indianapolis', *Antipode Monographs in Social Geography*, 1, 67-87

G.B., Department of Education (1967), 'Children and their Primary Schools: a Report', Report of the Committee Chaired by Lady Plowden, 2 vols. (London, HMSO)

G.B., Department of the Environment (1971a), 'Study of Rural Transport in Devon', Report of the steering committee

G.B., Department of the Environment (1971b), 'Study of Rural Transport in West Suffolk', Report of the steering committee

G.B., Department of Transport (1978a), 'Transport Act 1978, Public Transport Planning in Non-metropolitan Counties', Circular 8/78

G.B., Department of Transport (1978b), 'A Guide to Community Transport' (London, HMSO)

G.B., Department of Transport (1978c), 'Transport Statistics, Great Britain, 1966-76' (London, HMSO)

G.B., Department of Transport (1978d), 'Innovations in Rural Bus Services', Eighth Report from the Select Committee on Nationalised Industries, HC 635 (London, HMSO)

G.B., Department of Transport (1979), 'Concessionary Fares for Elderly, Blind and Disabled People', Cmnd 7475 (London, HMSO)

G.B., Department of Transport (1981), *Transport Statistics Great Britain 1970-1980* (London, HMSO)

Hägerstrand, T. (1970), 'What about People in Regional Science?' *Regional Science Association Papers*, 24, 7-21

Hägerstrand, T. (1972), 'The Impact of Social Organisation and Environment upon the Time Use of Individuals and Households', *Plan International Special Issue*, 24-30

Hansen, W.G. (1959), 'How Accessibility Shapes Land Use', *Journal of the American Institute of Planners*, 25, 73-6

Hillman, M., Henderson, I. and Whalley, A. (1973), 'Personal Mobility and Transport Policy', PEP Broadsheet 542, June

Hillman, M., Henderson, I. and Whalley, A. (1976), 'Transport Realities and Planning Policy', PEP Broadsheet 567

Hillman, M. and Whalley, A. (1979), 'Walking *is* Transport', PEP Broadsheet 583

Hodge, P. (1979), 'The Role of Rail in Rural Transport' in *Mobility in Rural*

Areas, Proceedings of a Conference held at Cranfield Institute of Technology, April, 137–47

Hopkin, J.M., Robson, P. and Town, S.W. (1978), 'The Mobility of Old People: a Study in Guildford', *Transport and Road Research Laboratory*, LR 850

Johnson, R.J. (1966), 'An Index of Accessibility and its Use in the Study of Bus Services and Settlement Patterns', *Tijdschrift voor Economische en Sociale Geografie*, 57 (1), 33–8

Joint Working Group of TRRL and Gwent County Council (1981), 'The Application of Accessibility Measures in Gwent; Travel to Hospitals and Shops', *Transport and Road Research Laboratory*, LR 994

Jones, S.R. (1981), 'Accessibility Measures: a Literature Review', *Transport and Road Research Laboratory*, LR 967

Koutsopoulos, K.C. and Schmidt, C.G. (1976), 'Mobility Constraints of the Carless', *Traffic Quarterly*, 30 (1), 67–83

Lenntorp, B. (1981), 'A Time Geographic Approach to Transport and Public Policy Planning' in Banister, D.J. and Hall, P.G. (eds.), *Transport and Public Policy Planning* (London, Mansell), 387–96

Ministry of Transport (1961), 'Rural Transport Services, Report of Preliminary Results from the Jack Committee' (London, HMSO)

Mitchell, C.G.B. and Town, S.W. (1977), 'Accessibility of Various Groups to Different Activities', *Transport and Road Research Laboratory*, SR 258

Morris, J.M., Dumble, P.L. and Wigan, M.R. (1979), 'Accessibility Indicators for Transport Planning', *Transportation Research*, 13A, 91–109

Moseley, M.J. (1979), *Accessibility: The Rural Challenge* London, Methuen)

Moseley, M.J. (1981), 'The Supply of Rural (In)accessibility', in Banister, D.J. and Hall, P.G. (eds.), *Transport and Public Policy Planning* (London, Mansell), 183–8

Moseley, M.J., Harman, R.G., Coles, O.B. and Spencer, M.B. (1977), *Rural Transport and Accessibility*, 2 vols. (University of East Anglia, Centre for East Anglian Studies)

Motor Agents Association (1981), personal communication

National Bus Company (1981), *Annual Report* (1980, NBC)

National Council for Voluntary Organisations (1980), 'Impact of New Technology on the Countryside', NCVO/Standing Conference of Rural Community Councils Futures Panel, October

Oldfield, R. (1979), 'Effect of Car Ownership on Bus Patronage', *Transport and Road Research Laboratory*, LR 872

Pahl, R.E. (1964), 'Urbs in Rure', London School of Economics and Political Science, Geographical Paper No. 2

Pickup, L. (1981), 'Housewives' Mobility and Travel Patterns', *Transport and Road Research Laboratory*, LR 971

Pirie, G.H. (1979), 'Measuring Accessibility: a Review and Proposal', *Environment and Planning A*, 11, 299–312

Randolph, W. and Robert, S. (1981), 'Population Redistribution in Great Britain, 1971–1981', *Town and Country Planning*, September, 227–31

Rhys, D.G. and Buxton, M.J. (1974), 'Car Ownership and the Rural Transport Problem', *Journal of the Chartered Institute of Transport*, 36, 109–12

Rigby, J. (1977), 'Access to Hospitals: a Literature Review', *Transport and Road Research Laboratory*, SR 853

Shaw, J.M. (ed.) (1979), *Rural Deprivation and Planning* (Geo Abstracts Ltd., University of East Anglia, Norwich)

Taylor, C. and Emerson, D. (1981), 'Rural Post Office: Retaining a Vital Service', Report by the National Council for Voluntary Organisations for the Development Commission

Thomas, C. and Winyard, S. (1979), 'Rural Incomes' in Shaw, J.M. (ed.), *Rural Deprivation and Planning* (Geo Abstracts Ltd., University of East Anglia, Norwich), 21–50

Town, S.W. (1980), 'The Social Distribution of Mobility and Travel Patterns', *Transport and Road Research Laboratory*, LR 948

Tunbridge, R.J. and Jackson, R.L. (1980), 'The Economics of Stage Carriage Operation by Private Bus and Coach Companies', *Transport and Road Research Laboratory*, LR 952

Tyson, W.J. (1972), 'The Peak in Road Passenger Transport: an Empirical Study', *Journal of Transport Economics and Policy*, 6 (1), 77–84

Wachs, M. and Kumagai, G. (1973), 'Physical Accessibility as a Social Indicator', *Socio-Economic Planning Sciences*, 7 (5), 437–56

Webster, F.V. (1977), 'Urban Passenger Transport: Some Trends and Prospects', *Transport and Road Research Laboratory*, LR 771

White, P.R. (1978), 'Midland Red's Market Analysis Project', *The Omnibus Magazine*, May/June, 61–8

7 RURAL COMMUNITIES

G.J. Lewis

Any study concerned with rural communities is, at the outset, bedevilled by problems of definition and the lack of an adequate conceptual framework. In a widespread and confused literature no agreement exists as to the nature of a rural community (Lewis, 1979), yet, for the purpose of discussion a general indication of how it is to be interpreted is necessary. In view of the decline of agricultural employment and the growth of commuterdom a pragmatic approach has to be adopted to the term '*rural*', such as that proposed by Wibberley (1972): 'it encompasses those parts of a country which show unmistakable signs of being dominated by extensive uses of land . . . this allows us to look at settlements which to the eye still appear to be rural but which, in practice, are mainly an extension of the city resulting from the development of the commuter train and the motor car.' As Hillery's (1955) wide-ranging review has revealed, a watertight definition of the term *community* is also impossible, although Reiss's (1959) summary provides a reasonable interpretation: 'a community arises through sharing a limited territorial space for residence and for sustenance, and functions to meet common needs generated in sharing this space by establishing characteristic forms of social action'. Within the rural environment, therefore, a community may involve a village, a hamlet, or even dispersed habitations; in other situations it can involve several settlements, and within some villages there may be more than one community.

Until recently those geographers and social scientists interested in the organisation and well-being of rural communities have lacked the concepts and methodologies of those involved in urban questions (Lewis, 1979). As a result most rural studies have tended to be empirical and holistic, thus lacking comparability and depth. Only of late has the rural geographer begun to develop some conceptual frameworks to guide his analysis and, as a consequence, a greater understanding of the complexity of the changing rural community has been achieved, which has placed the geographer in a better position to contribute more effectively to policy discussion.

Cycles of Interest

In attempting to review this field of enquiry it becomes readily apparent that geographers and other social scientists have over the years periodically shifted their main focus of interest. During the 1940s the main concern was with the distinctiveness of the 'rural way of life', which it was thought was best preserved in relatively isolated locations. Beginning with Arensberg's (1939) classic study in County Clare, Ireland, and replicated by Rees (1950) in north-east Montgomeryshire, Wales, a whole series of studies was carried out in the remoter parts of Western Europe, which emphasised the relationship between culture, economy and society within a recognisable locality (Davies and Rees, 1960; Franklin, 1969). Recently these studies have been ridiculed for their empiricism, which is rather unfair in view of the ideas that they contained and the impetus they provided for later research (Bell and Newby, 1971). During the next decade this theme was overtaken by a growing concern with the continued decline in rural employment opportunities and the consequent acceleration in rural depopulation (Saville, 1957; Zelinsky, 1962; Welsh Office, 1964; Merlin, 1971). By the 1960s the 'turnaround' in rural population had begun, particularly in those areas adjacent to major cities, which led to considerable interest being generated by what has become known as the 'metropolitan village'. For example, in Britain Elias and Scotson's (1965) pioneering study on Leicester's urban fringe stimulated considerable research activity into the social and spatial segregation of 'urban dominated' rural communities (Pahl, 1965; Masser and Stroud, 1965; Lewis, 1970).

Throughout these cycles of changing interest there was a clear realisation that the rural community was experiencing considerable social and economic hardships during the post-war period, a theme which was most evident when the causes and consequences of depopulation were discussed (President's Commission, 1967; Ashton and Long, 1972). Despite what the studies of the 1950s revealed as well as the actions of central government, local authorities and, in certain areas, development boards, it was not until the 1970s that the geographer appeared to become interested for the first time in questions involving welfare in general and deprivation in particular. To many this interest in rural deprivation reflects the changing focus of the discipline itself as well as the persistence of hardships despite public policies to alleviate them. However, to argue that the geographer has ignored the study of rural hardships until recently is to fail to appreciate the *raison d'être* of the changing foci of interest in the post-war period. Underlying all of these cycles of interest has been concern with two distinct, yet inter-related themes: social change and deprivation. What is significant and different today is that these two themes have themselves become a direct focus of attention.

Social Change

Any study of contemporary rural problems must begin by considering the nature of the changes experienced by rural communities during recent decades (Ford, 1978). The modernisation of agriculture and the urbanisation of rural society have wrought extensive changes in the countryside, and have been a source of considerable discussion over the years (Fuguitt, 1963; Williams, 1964). Traditionally, social change within the countryside has been associated with a rural-urban dichotomy or continuum, which emphasise differences in urban and rural values and behaviour (Frankenberg, 1966; Jones, 1973). As a result of the massive changes experienced by the countryside since the Second World War as well as the increasing links between town and country such a framework is of little relevance today. Therefore, in order to aid our understanding of contemporary rural change a framework is needed which is not only specifically rural, but also considers the rural community as part of a wider social and spatial system. Only of late have some tentative steps been taken to provide such a framework; for example, Lefaver (1978), admittedly within a planning sphere, emphasised the need for a separate rural context, yet ignored the dynamic nature of the rural-urban relationship, whilst Cloke (1978) highlighted such a relationship by conceiving the differentiation of rural communities within a distance-decay framework. Lewis and Maund (1976) took this view a stage further by adopting a more behavioural interpretation within a time-space framework. Rural social change was conceived as being initiated by three population movements, beginning with depopulation as a result of outward migration, thence population as a result of in-migration of young adventitious people and finally the repopulation of the countryside by those late in the family life-cycle. The operation of these movements in a time-space framework is illustrated in Figure 7.1. With the coming of industrialisation, with its demand for labour at specific locations and a fall in demand for rural labour, massive movements to the cities resulted, whilst at the same time some of the captains of industry, late

Figure 7.1: A Time-space Order of Urbanisation

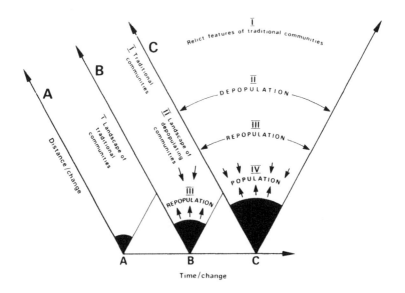

in their life-cycle, began to repopulate the countryside. At a later stage, the movement from the city to the countryside became an established feature, being particularly evident in those zones close to the city. But, it was also argued that all three movements occur together in all locations, and it is their relative proportions which vary from community to community, thus providing the basis for inter-community variation. For example, those communities experiencing a depopulation of their young might also be experiencing a repopulation by retired migrants as well as an influx of urban-dwellers seeking rural residence and employment. With the passage of time the latter process might become the predominant one. Therefore, it is the combination of these selective movements and their relative proportions which determine a community's character. Such an approach suggests that these differences within rural areas are of degree and not of kind, since urbanisation was conceived as a process affecting the whole of society.

The relevance of such a perspective is clearly exemplified by those recent studies which have attempted to analyse the patterns of rural society by means of a multivariate approach. At a national scale several studies have revealed differences with distance from major centres of population as well as distinctive regional variations (Christian, 1977;

Cloke, 1977, 1978; Webber and Craig, 1976; Webber, 1977). Probably the most detailed of these studies was that of Calmès and his associates (1978), who identified five types of French countryside: the Mediterranean coast, the agricultural west, the peri-urban areas of the north and north-east as well as the hinterlands of Paris and Lyons, plus two intermediate type regions. Similar patterns of differentiation have been revealed at regional and county scales (Lewis, 1979; Keys and McCracken, 1981), where socio-economic status, life-cycle and mobility were major factors of differentiation.

During the 1970s in the United States the pattern of social change itself began to change as a result of a more widespread repopulation of all parts of the countryside (Beale, 1975; Morrison and Wheeler, 1976; McCarthy and Morrison, 1979; Brown and Wardwell, 1980). More recently a similar 'turnaround' in rural population is apparent in Australia, Canada, Britain and a number of other developed countries (Australian Government, 1977; Vining and Kontuly, 1978; Bourne and Simmons, 1979; Vining, Pallone and Plane, 1981; Champion, 1981). According to Beale (1977) such a 'turnaround' is of utmost significance since it is 'occurring in hundreds of counties that conventional analysis in the 1960s would have consigned to continued stagnation and decline'. Even the remoter parts of Brittany and the Camargue in France (Picon, 1978; Eizner, 1978) and the Highlands of Scotland, mid-Wales and Cornwall (Jones, 1976; Lewis, 1981; Jones, 1981), are now being perceived as highly desirable for residence and retirement. According to Dillman (1979) this 'turnaround' involves several identifiable trends:

(1)   the continued suburbanisation of rural communities adjacent to metropolitan centres;
(2)   some decentralisation of manufacturing seeking lower land and wage costs;
(3)   acceleration of retirement migration into the countryside particularly as a result of early retirement;
(4)   a levelling off in the loss of agricultural population;
(5)   the exploitation of mineral and energy resources;
(6)   a desire for an alternative to urban life-style.

It would appear that the majority of these movers are motivated by non-economic factors (Zuiches and Rieger, 1978; Williams and Sofranko, 1979; Long and De Are, 1980), in particular a greater preference for rural life-styles and leisure. The continuing influx of urban commuters is increasingly being motivated by such factors as are the retirees (Allon-Smith, 1978) and the so-called urban dropouts (Schwarzwellar, 1979). These 'dropouts' are of significance since they include not only return migrants but also those who have deliberately forsaken urban employment as well as a growing number of professional and managerial people who work from home (Nicholson, 1975; Forsythe, 1980; Foeken, 1980). Inevitably this repopulation of the countryside takes place in a variety of guises including the development of small suburban-type estates within villages, the gentrification of existing housing, and the take-over of small farms for rural retreating (Gasson, 1967; McQuin, 1978). Of course, an integral part of this invasion of the countryside by the predominantly urban, middle classes is the acquisition of second homes, which for some is the first step towards permanent rural residence (Clout, 1974; Coppock, 1977).

This 'turnaround' in rural population must not, however, be over-emphasised, since depopulation is still rife in large parts of Europe (Council of Europe, 1980), and even in some relatively prosperous rural regions it still remains endemic (H.M. Treasury, 1976; Drudy, 1978). As in the past, contemporary depopulation is predominantly motivated by economic factors, as several recent regression-type analyses have revealed (Kriesberg and Vining, 1978; Todd, 1980, 1981), although in many areas the continuing loss of the young has by now become a 'way of life' (Hannan, 1970). Apart from these regional shifts of population local movements should also not be ignored since they can often contribute significantly to the growth or decline of a community's population (Lumb, 1980; White, 1981).

In any consideration of rural social change, however, the crucial element is not population growth or decline *per se* but rather the selectivity of people within each migratory flow (Wilkinson, 1978). Accordingly, Sherwood (1982) has argued that population turnover provides a more meaningful starting point in any consideration of social change since it identifies the demographic stability of a community as well as the character of the population flows involved. For example, Figure 7.2 reveals the pattern of population turnover in south-west Northamptonshire, which suggests broad spatial variations within the context of competing urban centres, tenure structures and planning policies.

In order to gain some understanding of the role of selective migrations in generating changes within communities it is necessary to focus on what actually changes, how the change takes place, and the resultant diversity of communities. According to Lewis and Maund

Figure 7.2: Household Turnover in South-west Northamptonshire Parishes, 1975–80

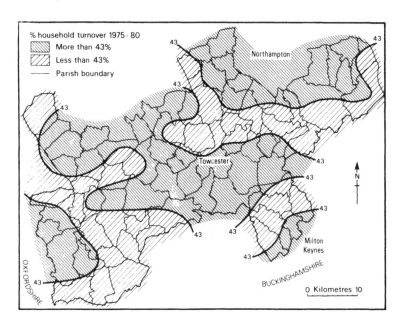

(1976) change within a community may be considered in terms of three inter-related components: the social structure of its residents, their value system, and behaviour patterns. From Figure 7.3 it can be seen that each component can be affected differently by each of the three major forms of population movements. To many the basis of the social structure of any community is that of social class; following Pahl's (1965) study in Hertfordshire numerous studies in other urban fringes have confirmed this claim (Ambrose, 1974; Connell, 1978; Pacione, 1980). Essentially, what results is the creation of two communities within one place, each involving separate worlds within and beyond the locality. On the other hand, where depopulation predominates then the whole fabric of the community collapses, since a small and ageing population find it impossible to sustain a viable set of local activities, thus reducing social interaction to only that of a primary nature. Of course, this emphasises a continuation of a one class community (Mitchell, 1950; Brody, 1974; Commins, 1978). As a result of such evidence there has been a tendency to characterise the rural community as 'metropolitan' and 'residual', or 'encapsulated' and 'occupational'

Figure 7.3: The Components of an Urbanisation System Within the Countryside

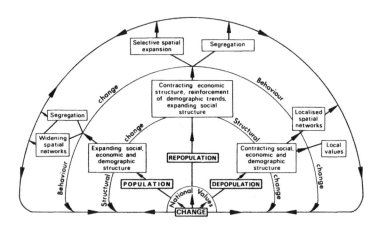

(Newby, 1980), which appears a rather simplistic and restrictive interpretation of rural social change. The spread of the 'turnaround' in population throughout the countryside and the increasing use of more analytical methods within rural geography has led to the questioning of the validity of these typologies. For example, several recent studies have questioned the usefulness of social class as a means of interpreting behaviour within communities. Often villages in close proximity with relatively similar social structures can have different patterns of social interaction and involvement in community organisations (Jackson, 1968; Radford, 1970). What is evident is that the middle-class rural dwellers are not a homogeneous group. Some have chosen to reside in the countryside ('community' migrants), whilst others have been 'forced' out of the city by housing opportunities ('reluctant' migrants). This suggests that life-style may be a more significant basis for community development (Walker, 1977; Everitt, 1980). Similarly, Lewis and Maund (1976) and Palmer, Robinson and Thomas (1977) have shown that attitudes towards various facets of rural living are related more to the degree of community involvement than to social class and length of residence. It would appear that even though the middle classes and working classes live in different worlds outside the community, within it 'local' interests often lead to the interaction and involvement of parts of the two classes. Such developments are more apparent in smaller than larger places, and become particularly

Figure 7.4: Arson Attacks on Second-home Residences in Wales, Reported 1978-80

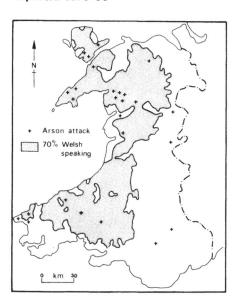

evident when the community is under a threat of change (Byron and MacFarlane, 1980).

Apart from the growth of commuterdom, the only other process involved in the recent population 'turnaround' to have claimed widespread attention has been the emergence in a number of developed countries of second-home residences (Coppock, 1977). During the past decade these residences have generated much controversy as to their impact and benefit for the communities involved. For example, Davies and O'Farrell (1981) have shown that second homes predominate within villages rather than among dispersed habitations, which results in considerable dislocation within local housing markets (de Vane, 1975) and reduces the viability of communities (Bollom, 1978; Sarre, 1981). In Wales these issues, combined with nationalistic feelings, have resulted in widespread arson attacks on second-home cottages (Figure 7.4).

In any discussion of the changing rural community consideration must, however, be given to the impact of public policy towards population movements. For example, in England and Wales, a 'key' settlement policy which concentrates development and population into a limited number of central villages (Cloke, 1978), not only 'guides' local movements but also influences the social composition of communities. These villages designated as 'non-growth' have become via the gentrification process increasingly attractive to the upper middle class whilst developments in 'key' villages have primarily been for 'suburbanites' (Herrington and Evans, 1979–80; Martin and Voorhees Associates, 1981). This often results in the existence of two socially different communities within close proximity to each other. There is considerable doubt, therefore, whether this policy achieves what Groot (1979) regards as the prime purpose of any settlement planning; the continuity of local communities by the survival of local community generating activities such as church, school, club and societies. It is to the consideration of these and other aspects of 'quality of life' in the countryside that we must now turn.

Rural Deprivation

The long-term effect of the changing employment and population structure of rural communities has been to accentuate the 'difficulties' being experienced by their residents (Henderson Report, 1974; Countryside Review Committee, 1976; Carter Administration, 1979). Such a theme is not a new one since rural people have always suffered hardships of some kind or another. Since the Second World War diminishing employment opportunities, depopulation, and increasing personal mobility has resulted in a decline in service provision. In turn this reduced demand has led to further out-migration in search of better conditions and employment, which in turn means a further decrease in demand and hence provision (Countryside Review Committee, 1977; Ferge, 1980). During the last three decades national and local governments have attempted to improve the fortunes of the countryside by means of specific agricultural policies as well as those of a more general nature (Sadler and Mackay, 1977). In Scotland and Wales local development boards have been instituted in order to aid the more marginal rural regions. Of course, during the same period large-scale developments, such as the growth in the holiday trade, expansion of forestry, the construction of reservoirs, hydro-electric schemes and nuclear power stations, and in the case of north-east Scotland, oil developments, have had a significant impact upon community life and the socio-economic conditions of the areas involved (Sadler and Mackay, 1977; Sewel, 1979). Some researchers believe that these economic changes have led to marked improvements in the conditions of the countryside, thus providing a basis for a genuine rural recovery (Glyn-Jones, 1979; Wenger, 1980). Such a recovery must, however, not be overemphasised since insufficient employment, low incomes, low female activity rates, decline of basic services such as shops, doctors, chemists and schools, inadequate housing provision, poor accessibility to many services, and low levels of personal mobility still abound (Shaw, 1979; Walker, 1979). Since the Second World War in many fields of policy and provision, there has been a failure to adjust to the specific needs of rural communities whilst at the same time rural residents have little control over decisions affecting their lives (Maddox, 1973; Gilder, 1979; Avery, Lonsdale and Volgyes, 1980).

However, as Shaw (1979) has rightly noted, the increased concern with rural social problems 'has not been matched by a systematic appraisal of the nature and extent of the social difficulties experienced in many rural areas'. Despite the well-known pitfall of attempting to define the undefinable and measure the unmeasurable, several approaches have been developed to determine the nature of contemporary rural deprivation. At a regional scale the 'spatial inequality' approach has been very much in favour, it generally involving a multivariate analysis of a set of 'deprivation' variables (Gordon and Whittaker, 1972; Ross, Bluestone and Hines, 1979; Bracken, 1980; Knox and Cottam, 1981a, b). This approach has revealed quite clearly the existence of marked spatial variations in the incidence of rural deprivation; for example, Figure 7.5 suggests that in a part of the highlands of Scotland deprivation is concentrated at the two ends of the settlement hierarchy, with intermediate areas characterised by various distinctive syndromes of deprivation (Knox and Cottam, 1981a, b). Such evidence tends to negate the rural sociologist's habit of conceiving a region as a single entity and so falling into the trap of oversimplification and over-generalisation (Carter, 1974). At an individual level Groot has introduced what has become known as the 'livability' approach, arguing that 'the continuing or reaching of a certain population or a minimum level of services or a physical residential climate that satisfies certain demands do not matter, but what matters is the staying or not of a local system in which one feels well' (Groot, 1979). In this approach the immaterial aspects of life are considered of greater significance than the material and, of course, it could be argued that livability and viability are primarily of an immaterial character. Those studies which have used 'quality of life' indices have themselves emphasised the necessity of considering immaterial aspects in any discussion of rural

Figure 7.5: Deprivation in a Part of the Highlands of Scotland

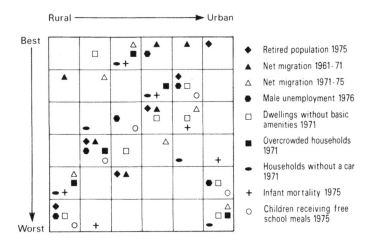

deprivation (Rojek, Clemente and Summers, 1955; Dillman and Tremblay, 1977; Knox and Cottam, 1981a, 1981b). This may be illustrated from Pacione's (1980) study of a small community on Glasgow's urban fringe where, apart from employment, the actual life concerns of the middle and working classes differed quite markedly; the former emphasised their own home and the village as a place to live, whilst the latter were more concerned with service provision and village change. However, these studies suggest that the immaterial aspects can only really be considered if conceived within some compensatory model. Such a model assumes that drawbacks are at least compensated for by advantages in a framework of certain margins and tolerations, and the existence of individuals with similar concepts and weightings of advantages and disadvantages. Van Bemmel (1981) has developed such a model, based on Clark's 'stress-resistance' model of migration potential, in a study of deprivation in the remoter parts of Zuidwest Friesland in the Netherlands. Clearly, irrespective of whatever approach is adopted it has been revealed that the countryside as a whole experiences deprivation, as well as some advantage, relative to urban areas, and that some localities are more disadvantaged than others. In addition it has been emphasised that certain groups, such as the unemployed, poor, elderly, car-less, young housewives, and so on, experience greater deprivation than others living in the same locality, illustrating once again that those in greatest need invariably experience the greatest difficulty in meeting those needs. With the continuing decline in employment opportunities and service provision, inevitably accessibility has become the root cause of rural disadvantage (Moseley, 1979; Veldman, 1981).

Of the various components which comprise the rural 'quality of life' matrix three have been extensively researched – the decline in job opportunities (Lucey and Kaldor, 1969; Sewel, 1979; Moore, 1981; Hodge and Whitby, 1981), increasing competition and conflict in the housing market (Gasson, 1975; Shucksmith, 1981; Dunn, Rawson and Rogers, 1981), and the continuing decline in public transport (Moseley, 1979). These are considered elsewhere in this volume. Another major element in a satisfactory quality of life is the availability of an adequate service provision, a theme which is only beginning to be considered in any meaningful way. Since the Second World War there has been a continuous decline of rural services, whether it be retailing, chemists, doctors, schools, chapels, and so on, which has been explained in terms of a falling population and a trend for provision to be centralised in larger places (Thomas, 1972; Johansen and Fuguitt, 1973). However, there is growing evidence to suggest that this decline continues even though the population may be growing, which suggests that multi-purpose journeys characterise much of rural consumer movements (Standing Conference of Rural Community Councils, 1978). For the benefit of the immobile groups 'key' settlement policy in Britain has attempted to encourage the localisation of services in 'central' villages, yet, the decline of services is taking place in both 'key' and 'non-key' villages, thus, indicating and overall upward shift in the service hierarchy. In order to produce a more effective policy a greater understanding is required of the spatial manifestation of service decline (Weekley, 1977), and the motivations underlying the consumer movements of various rural groups (Holmes and Brown, 1978). In this context many of the earlier central-place type studies could provide a useful initial framework of analysis and some guidelines for suggesting alternative solutions (Rowley, 1971). Recently it has been shown that whatever solution is proposed it will have to overcome the social and spatial biases in information access which characterised rural dwellers (Busch, 1978; Clark amd Unwin, 1980).

A service which is causing increasing concern within the countryside is that of health care, which even in Britain with its nationalised service still has significant inequalities (Haynes and Bentham, 1979;

Phillips, 1980). For example, despite attempts to determine the distribution of general practitioners by population numbers the areal units employed are so large that Hart's 'inverse care law' is still evident (Gibson and Whitelegg, 1979; Williams, Bloor, Horobin and Taylor, 1980). A recent survey in a part of rural Hampshire has revealed that the tendency for practices to amalgamate into larger units with a greater range of expertise and facilities has resulted in an increasing proportion of the population residing more than 10 miles from a surgery, and nearly 25 per cent have to wait over an hour for a return bus (Winchester and Central Hampshire Community Health Council, 1981). Phillips (1979) has confirmed that the journey to visit a surgery is a greater problem for the less mobile lower status group than for the higher status group, although the Budds (1981) suggest that it is the elderly and the young mothers who face the greatest difficulties. This evidence has important implications in the planning of further primary medical care in the countryside because it emphasises the need for greater home visiting if patients cannot travel to a surgery. In those countries where a national health service is more limited these problems are even more acute (Girt, 1973; Walmsley and McPhail, 1976; Walmsley, 1978).

Up to the present few studies have analysed rural deprivation from the viewpoint of the various deprived groups (Moseley, 1978; Gruer, 1975). An interesting example of such an approach was Nokes's (1982) study of the behaviour and attitudes of the elderly, including long-term residents and retirement migrants, in a remote corner of lowland England. The overwhelming majority experienced some difficulties in gaining access to shops, medical care, and social services, yet were relatively happy with their housing conditions, had plenty of friends, and were highly involved in community activities. Only a small minority had considered leaving the district and, on balance, the majority were satisfied with their present situation. It would appear that the long-term residents had adjusted to rural conditions over several decades, whilst the more recent retirees in choosing rural residence were aware of both its advantages and disadvantages. The evidence contained in this study suggests the need for a greater use of a compensatory approach in any consideration of rural deprivation.

In a geographical investigation of rural deprivation several significant issues are raised, of which two are worthy of further consideration. First, to what extent are these 'hardships' specifically rural in their character? Many would argue that they have little to do with space, rather they are due to the underlying structural forces relating to our

Figure 7.6: Overlapping Sets of Problems in Urban and Rural Areas

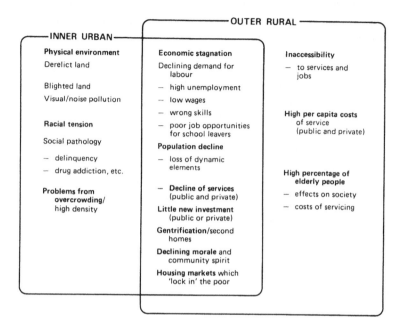

economy and society as a whole. Therefore, rural deprivation is inextricably linked to that of urban deprivation and, according to Moseley (1980), both involve high unemployment, low wages and limited job opportunities, which leads to out-migration of the more skilled, and results in those less able to compete, such as the poor, elderly, unskilled, and so on, being left behind, usually in old and run-down housing. In such circumstances, private and public services cannot survive, resulting in another turn in the cycle of deprivation. However, where rural and urban deprivation differ is in their environmental and spatial manifestations. Within the city deprivation is related to overcrowding, poor housing, ethnic relations, and social pathology (Thrift, 1979); on the other hand, rural deprivation involves inaccessibility caused by the spatial configuration of its population and settlement (Figure 7.6). Those 'in need' are much more dispersed and isolated in their distribution in the countryside than they are in the city and this makes the provision of even basic services extremely expensive (Nutley, 1980). A declining and dispersed population has resulted in a lack of a threshold population sufficient to sustain the most basic services and facilities. The elderly, unemployed, car-less, etc, therefore find it increasingly difficult to reach even those services needed for everyday life. Within the countryside 'space' is a vital parameter, which suggests that in any consideration of rural deprivation it is necessary to distinguish between 'place' and 'people' deprivation (Kirby, 1981; Veldman, 1981). The second issue raises the question, to what extent does deprivation manifest social change? Certainly the effect of deprivation has been to accelerate the out-migration of the native population, thus leading to social isolation and a break-up of community life (Lewis, 1979). Similarly, the closure of schools is generally felt to have a direct effect on the number of young families who will choose to remain or move into these settlements, bringing hope of maintaining a balanced community. On the other hand, the entry of older, middle-class people can often provide the 'power' necessary to voice the 'hardships' felt at both local and national level, as evidenced in Britain by the now rejected Education (No. 2) Bill, with its proposal to allow county councils to charge parents for bussing country children to school.

Rural Future

In the 1980s those geographers interested in the changing rural community and its attendant problems should begin to be concerned with the rural community of the future (Köppä, 1981). In the 1960s the countryside witnessed the coming of the urban commuter; in the 1980s and beyond what is the life-style of the rural dweller going to be like? If the micro-chip revolution continues to shed labour and provides potential for industrial and administrative activities to be operated at a distance then the population 'turnaround' is likely to accelerate. On the other hand, a severe energy crisis could well attenuate rural population growth and localise economic and social activities. Of course, both scenarios could occur simultaneously. Yet, whichever one favours, rural life-styles are going to change, and the countryside is going to be a very different place (Coughenor and Busch, 1978; Bradshaw and Blakesley, 1979). The implication of these changes for those already in the countryside as well as the 'newcomers' cannot be ignored by the geographer.

In order to achieve a deeper understanding of the contemporary and likely future rural community geographers must move a step further up the ladder of explanation, particularly by focusing upon those who 'control' the allocation of resources whether it be housing, employment, service provision, transport, energy, and so on. In other words, who decides, who pays, and who benefits? Such a perspective involves both local and external agents of control. Recently a number of rural sociologists have pointed the way towards a better understanding of rural life by concentrating on the actions of such powerful groups as landowners, farmers and the middle-class newcomers (Newby, Bell, Rose and Saunders, 1978). What is suggested is that these groups, usually in concert but sometimes in opposition, further their interest, often at the expense of the 'locals', through involvement in local authority decision-making. With the emergence of pressure groups, however, such a distinction becomes less relevant, as newcomers and locals often combine to defend their community (Lowe, 1977; Leat, 1981). For a more complete picture it is necessary to look beyond the locality and consider the actions of government departments, county and district councils, regional health and water authorities, financial institutions, planning authorities, and so on. Even in national politics the rural lobby must be identified and understood; for example, in Australia a rural-based party forms a part of a governing coalition whilst in Britain the recently formed association (Rural Voice) of eight bodies with rural interests is already having some influence on the rural policies of several national political parties. Among the issues which the geographer needs to consider about these agencies include: how do they operate? do they have a specific policy towards rural areas? how do they perceive rural space and society? what are the constraints operating on these agencies in the allocative process? and what is the impact of their decisions on the life chances of rural dwellers? Several recent sociological essays have taken these issues a stage further by advocating the adoption of a more critical perspective (Newby, 1978; Buttel and Newby, 1980); however, the relevance of this perspective to rural living will ultimately depend more upon its contribution to the creation of successful rural policies than to its acceptance by fellow rural sociologists. Only by analysis and action rather than description and critique, can the geographer contribute more effectively to an understanding of the problems of living in a rapidly changing rural community.

References

Allon-Smith, R.D. (1978), *The Migration of the Elderly: A Social Geography*, unpublished PhD thesis, University of Leicester

Ambrose, P. (1974), *The Quiet Revolution* (Chatto and Windus, London)

Arensberg, C.M. (1939), *The Irish Countryman. An Anthropological Study* (Macmillan, New York)

Ashton, J. and Long, W.H. (eds.) (1972), *The Remoter Rural Areas of Britain* (Oliver and Boyd, Edinburgh)

Australian Government (1977), *Trends in Non-metropolitan Population Growth and their Relevance to Decentralisation Policy*, Decentralisation Policy Branch, Department of the Environment, Housing and Community Development, Canberra, draft

Avery, W.P., Lonsdale, R.E. and Volgyes, I. (1980), *Rural Change and Public Policy: Eastern Europe, Latin America and Australia* (Pergamon, New York)

Beale, C.L. (1975), *The Revival of Population Growth in Non-Metropolitan America*, Economic Research Service, O.S. Department of Agriculture, ERS-605

Beale, C.I. (1977), 'The Recent Shift of United States Population to Non-metropolitan Areas, 1970-75', *International Regional Science Review*, 2, 113-22

Bell, C. and Newby, H. (1971), *Community Studies* (George Allen and Unwin, London)

Bollom, C. (1978), *Attitudes and Second Homes in Rural Wales*, Social Science Research Monograph No. 3 (University of Wales, Board of Celtic Studies, Cardiff)

Bourne, L.S. and Simmons, J.W. (1979), *Canadian Settlement Trends: An Examination of the Spatial Pattern of Growth, 1971-76*, Centre for Urban and Community Studies, Major Report 15, University of Toronto

Bracken, I. (1980), 'Socio-economic Profiles for Rural Areas: a Study in Mid-Wales', *Cambria*, 7, 29-44

Bradshaw, K. and Blakesley, E.J. (1979), *Rural Communities in Advanced Industrial Society* (Praeger, New York)

Brody, H. (1974), *Irish Killane: Change and Decline in the West of Ireland* (Penguin, Harmondsworth)

Brown, D.L. and Wardwell, J.W. (eds.) (1980), *New Directions in Urban-Rural Migration: The Population Turnaround in Rural America* (Academic Press, New York)

Budd, A.J. and J.E.S. (1981), 'Politics, Health and Mass Media: a Geographical Perspective' in D. Turnock (ed.), *Leicestershire*, Geographical Essays, University of Leicester Geography Department, Occasional Paper 1, 19-36

Busch, L. (1978), 'On Understanding Understanding. Two views of Communication', *Rural Sociology*, 43, 450-73

Buttel, F.H. and Newby, H. (eds.) (1980), *The Rural Sociology of the Advanced Societies* (Croom Helm, London)

Byron, R. and MacFarlane, G. (1980), *Social Change in Dunrossness: a Shetland Study*, North Sea Oil Panel Occasional Paper No. 1 (Social Science Research Council, London)

Calmès, R. (1978), *L'Espace Rural Francais* (Masson, Paris)

Carter Administration (1979), *The Small Community and Rural Development Policy* (US Government Printing Office, Washington)

Carter, I. (1974), 'The Highlands of Scotland as an Underdeveloped Region' in de Kadt, E. and Williams, G. (eds.), *Sociology of Development* (Tavistock, London), 131-45

Champion, A.G. (1981), *Counter-urbanisation and Rural Rejuvenation in Britain: An Evaluation of Population Trends since 1971*, Department of Geography, Seminar Paper No. 38, University of Newcastle-upon-Tyne

Christian, C. (1977), 'La typologie de l'agriculture de Belgique', *Travaux Géographiques de Liège*, 165, 93-109

Clark, D. and Unwin, K.I. (1980), *Information Services in Rural Areas* (Geo Books, Norwich)

Cloke, P.J. (1977), 'An Index of Rurality for England and Wales', *Regional Studies*, 11, 31-46

Cloke, P.J. (1978), 'Changing Patterns of Urbanisation in Rural Areas of England and Wales, 1961-1971', *Regional Studies*, 12, 603-17

Cloke, P.J. (1979), *Key Settlements in Rural Areas* (Methuen, London)

Clout, H.C. (1974), 'The Growth of Second-home Ownership: an Example of Seasonal Suburbanization' in J.H. Johnson (ed.), *Suburban Growth, Geographical Processes at the Edge of the Western City* (Wiley, London), 101-27

Commins, P. (1978), 'Socio-economic Adjustments to Rural Depopulation', *Regional Studies*, 12

Connell, J. (1978), *The End of Tradition – Country Life in Central Surrey* (Routledge and Kegan Paul, London)

Coppock, J.T. (ed.) (1977), *Second Homes: Curse or Blessing* (Pergamon, Oxford)

Coughenor, C.M. and Busch, L. (1978), 'Alternative Futures for Rural America: the Cloudy Crystal Ball', in T.R. Ford (ed.), 211, 246

Council of Europe (1980), *Methods to Stop Rural Depopulation and to Involve Citizens in the Development of their Regions* (Strasbourg)

Countryside Review Committee (1976), *The Countryside – Problems and Policies*, A Discussion Paper (Department of the Environment, London)

Countryside Review Committee (1977), *Rural Communities*, Topic Paper No. 1 (Department of the Environment, London)

Davies, E. and Rees, A.D. (eds.) (1960) *Welsh Rural Communities* (University of Wales Press, Cardiff)

Davies, R.B. and O'Farrell, P.N. (1981), *An Intra-regional Locational Analysis of Second Home Ownership*, Papers in Planning Research No. 23 (Department of Town Planning, University of Wales, Institute of Science and Technology)

De Vane, R. (1975), *Second Home Ownership: A Case Study*, Bangor Occasional Papers in Economics, No. 6 (University of Wales, Cardiff)

Dillman, D.A. (1979), 'Residential Preferences, Quality of Life, and the Population Turnaround', *American Journal of Agricultural Economics*, 61, 960-6

Dillman, D.A. and Tremblay, K.R. (1977), 'The Quality of Life in Rural America', *The Annals of the American Academy of Political and Social Science*, 429, 115-29

Drudy, P.J. (1978), 'Depopulation in a Prosperous Agricultural Subregion', *Regional Studies*, 12 49-60

Dunn, M., Rawson, M. and Rogers, A. (1981), *Rural Housing: Competition and Choice* (George Allen and Unwin, London)

Eizner, N. (1978), 'Les landes de Lanvaux aujourd'hui. Une voie novelle pour socialisation de l'agriculture privée', *Annales de Geographie*, 88, 351-68

Elias, N. and Scotson, J.L. (1965), *The Established and the Outsiders* (Cass, London)

Everitt, J. (1980), 'Social Space and Group Life-styles in Rural Manitoba', *Canadian Geographer*, 24, 237-54

Ferge, Z. (1980), 'Dynamics of Deprivation', Paper prepared for European Centre for Social Welfare Training and Research (Vienna)

Foeken, D. (1980), 'Return Migration to a Marginal Rural Area in North-western Ireland', *Tijdschrift voor Economische en Sociale Geografie*, 71, 114-20

Ford, T.R. (1978), *Rural U.S.A.: Persistence and Change* (Iowa State University Press, Ames)

Forsythe, D.E. (1980), 'Urban Incomers and Rural Change: the Impact of
    Migrants from the City on Life in an Orkney Community', *Sociologia Ruralis*,
    20, 287–307
Frankenberg, R. (1966), *Communities in Britain* (Penguin, Harmondsworth)
Franklin, S.H. (1969), *The European Peasantry: The Final Phase* (Methuen,
    London)
Fuguitt, G.V. (1963), 'The City and the Countryside', *Rural Sociology*, 28,
    246–61
Gasson, R. (1967), 'Some Economic Characteristics of Part-time Farming in
    Britain', *Journal of Agricultural Economics*, 18, 111–20
Gasson, R. (1975), *Provision of Tied Cottages*, Department of Land Economy,
    Occasional Paper No. 4 (University of Cambridge)
Gibson, C.T. and Whitelegg, J. (1979), 'Access to Health Care Facilities in
    Cumbia, in Halsall, D.A. and Turton, B.J. (eds.), *Rural Transport Problems
    in Britain*, Transport Geography Study Group, Institute of British
    Geographers
Gilder, I.M. (1979), 'Rural Planning Policies: an Economic Appraisal', *Progress
    in Planning*, 11 (3), 213–71
Girt, J.L. (1973), 'Distance to General Medical Practice and its Effect on Revealed
    Ill-health in a Rural Environment', *The Canadian Geographer*, 17, 154–66
Glyn-Jones, A. (1979), *Rural Recovery: Has it Begun* (Devon County Council and
    the University of Exeter, Exeter)
Gordon, L. and Whittaker, R.M. (1972), 'Indicators of Local Prosperity in the
    South West Region', *Regional Studies*, 6, 299–313
Groot, J.P. (1979), 'Kleine rurale Kernen' in F. Grunfeld (ed.), *Gebouwde
    Omgeving – neerslag ven onze samenleving* (Alphen ald, 120–42)
Gruer, R. (1975), *The Needs of the Elderly in the Scottish Borders* (HMSO,
    Edinburgh)
Hannan, D. (1970), *Rural Exodus* (Chapman, London)
Haynes, R.M. and Bentham, C.G. (1979), *Community Hospitals and Rural
    Accessibility* (Teakfield, Farnborough)
Henderson Report (1974), *Rural Poverty in Northern New South Wales*
    (Australian Government Commission of Inquiry into Poverty, Canberra)
Herrington, J. and Evans, D. (1979-80), *Household Movements and Planning
    Policy*, Working Papers 1–5 (Department of Geography, Loughborough
    University)
Hillery, G.A. (1955), 'Definitions of Community: Areas of Agreement', *Rural
    Sociology*, 20, 111–23
H.M. Treasury (1976), *Rural Depopulation: Report by an Interdepartmental
    Group* (London, HMSO)
Hodge, I. and Whitby, M. (1981), *Rural Employment* (Methuen, London)
Holmes, J.H. and Brown, J. (1978), 'Travel Behaviour and Contact Systems
    of Isolated Rural Populations in Southern Queensland', Paper presented at the
    Joint Australian-United States Seminar on Problems of Sparsely Populated
    Areas, Adelaide
Jackson, V.J. (1968), *Population in the Countryside: Growth and Stagnation in
    the Cotswolds*, West Midlands Social and Political Research Unit, University
    of Birmingham
Johansen, H.E. and Fuguitt, G.V. (1973), 'Changing Retail Activity in Wisconsin
    Villages', *Rural Sociology*, 57, 267–300
Jones, G.E. (1973), *Rural Life* (Longman, London)
Jones, H.R. (1976), 'The Structure of the Migration Process: Findings from a
    Growth Point in Mid-Wales', *Transactions of the Institute of British
    Geographers*, N.S. 2 (4), 421–32

Jones, H.R. (1981), *Recent Migration in Northern Scotland: Pattern, Process, Impact*, SSRC, North Sea Oil Panel Occasional Paper No. 8

Keys, C.L. and McCracken, W.J. (1981), 'An Ecological Analysis of Demographic Variation in Mid New South Wales', *Australian Geographer*, 15, 27-38

Kirby, A. (1981), 'Geographic Contribution to the Inner City Deprivation Debate: a Critical Assessment', *Area*, 13, 177-81

Knox, P.L. and Cottam, M.B. (1981a), 'A Welfare Approach to Rural Geography: Contrasting Perspectives on the Quality of Highland Life', *Transactions of the Institute of British Geographers*, N.S.6, 433-50

Knox, P.L. and Cottam, M.B. (1981b), 'Rural Deprivation in Scotland: a Preliminary Assessment', *Tijdschrift voor Economische en Sociale Geografie*, 72, 162-75

Köppä, T. (1981), *How Rural is our Future?* Abstracts of Papers, Eleventh European Congress for Rural Sociology, Helsinki

Kriesberg, E.M. and Vining, D.R. (1978), 'On the Contribution to Changes in Net Migration: a Time Series Confirmation of Beale's Cross-sectional Results', *Annals of Regional Science*, 12, 1-11

Leat, D. (1981), 'The Role of Pressure Groups in Rural Planning', *Countryside Planning Yearbook 1981* (Geo Abstracts, Norwich), 71-83

Lefaver, S. (1978), 'A New Framework for Rural Planning', *Urban Land*, 37, 7-13

Lewis, G.J. (1970), 'Suburbanisation in Rural Wales: a Case Study' in Carter, H. and Davies, W.K.D. (eds.), *Urban Essays: Studies in the Geography of Wales* (Longman, London), 144-76

Lewis, G.J. (1979), *Rural Communities. A Social Geography* (David and Charles, Newton Abbott)

Lewis, G.J. (1981), 'The Repopulation of the Welsh Countryside: Patterns and Implications', unpublished paper, Department of Geography, University of Leicester

Lewis, G.J. and Maund, D.J. (1976), 'The Urbanisation of the Countryside: a Framework for Analysis', *Geografiska Annaler B*, 58, 17-27

Long, L.H. and De Are (1980), *Migration to Non-metropolitan Areas: Appraising the Trend and Reasons for Moving*, US Bureau of the Census, Special Demographic Analysis, CDS 80-2

Lowe, P.D. (1977), 'Amenity and Equity: a Review of Local Environmental Pressure Groups in Britain', *Environment and Planning*, 49, 35-55

Lucey, D.I.F. and Kaldor, D.R. (1969), *Rural Industrialization: The Impact of Industrialization on Two Rural Communities in Western Ireland* (Geoffrey Chapman, London)

Lumb, R. (1980), *Migration in the Highlands and Islands of Scotland*, Institute for the Study of Sparsely Populated Areas, University of Aberdeen, Research Report 3

McCarthy, K.F. and Morrison, P.A. (1979), *The Changing Demographic and Economic Structure of Non-metropolitan Areas in the United States*, Rand Corporation, Santa Monica, R-2399-EDA

McQuin, P. (1978), *Rural Retreating: A Review and an Australian Case Study* (Department of Geography, University of New England, Armidale)

Maddox, J.G. (1973), *Toward a Rural Development Policy* (National Planning Association, Washington)

Martin and Voorhees Associates (1981), *Rural Settlement Policy in England and Wales* (London)

Masser, E.I. and Stroud, D.C. (1965), 'The Metropolitan Village', *Town Planning Review*, 36, 111-24

Maund, D.J. (1970), *The Urbanization of the Countryside: a Case Study in*

Herefordshire, unpublished MA thesis, University of Leicester

Merlin, P. (1971), *L'Exode Rural* (Presses Universitaires de France, Paris)

Mitchell, G.D. (1950), 'Depopulation and Rural Social Structure', *Sociological Review*, 42 (1950), 11-24

Moore, R. (ed.) (1981), *Labour Migration and Oil*, SSRC North Sea Oil Panel Occasional Paper 7

Morrison, P.A. and Wheeler, J.P. (1976), 'Rural Renaissance in America?' *Population Bulletin*, 31, 1-27

Moseley, M.J. (1978), *Social Issues in Rural Norfolk*, Centre of East Anglian Studies, University of East Anglia

Moseley, M.J. (1979), *Accessibility: The Rural Challenge* (Methuen, London)

Moseley, M.J. (1980), *Rural Development and its Relevance to the Inner City Debate* (Social Science Research Council, London)

Newby, H. (ed.) (1978), *International Perspectives in Rural Sociology* (Wiley, Chichester)

Newby, H. (1980), *Green and Pleasant Land?* (Penguin, Harmondsworth)

Newby, H., Bell, C., Rose, D. and Saunders, P. (1978), *Property, Paternalism and Power: Class and Control in Rural England* (Hutchinson, London)

Nicholson, B. (1975), 'Return Migration to a Marginal Rural Area – an Example from North Norway', *Sociologia Ruralis*, 15, 24-9

Nokes, J. (1982), *The Elderly in a Rural Environment: a Geographical Perspective*, unpublished PhD thesis, University of Leicester

Nutley, S.D. (1980), 'The Concept of Isolation', *Regional Studies*, 14, 111-24

Pacione, M. (1980), 'Differential Quality of Life in a Metropolitan Village', *Transactions of the Institute of British Geographers*, 5, 185-206

Pahl, R.E. (1965), *Urbs in Rure: The Metropolitan Fringe in Hertfordshire*, Geographical Paper No. 2, London School of Economics

Palmer, C.J., Robinson, M.E. and Thomas, R.W. (1977), 'The Countryside Image – an Investigation of Structure and Meaning', *Environment and Planning*, 9A, 739-50

Phillips, D.R. (1979), 'Spatial Variations in Attendance at General Practitioner Services', *Social Science and Medicine*, 13D, 169-81

Phillips, D.R. (1980), *Contemporary Issues in the Geography of Health Care* (Geobooks, Norwich)

Picon, B. (1978), 'Mécanismes sociaux de transformation d'un écosystème fragie: la Camargue', *Etudes Rurales*, 71-2, 219-29

President's Commission (1967), *Rural Poverty in the United States*, National Advisory Commission on Rural Poverty (Government Printing Office, Washington)

Radford, C. (1970), *The New Villagers: Urban Pressure on Rural Areas in Worcestershire* (Frank Cass, London)

Rees, A.D. (1950), *Life in a Welsh Countryside* (University of Wales Press, Cardiff)

Reiss, A.J. (1959), 'The Sociological Study of Communities', *Rural Sociology*, 24, 111-27

Rojek, D.G., Clemente, F. and Summers, G.F. (1975), 'Community Satisfaction: a Study of Contentment with Local Services', *Rural Sociology*, 40, 177-92

Ross, P.J., Bluestone, H. and Hines, F.K. (1979), *Indicators of Social Well-Being for U.S. Counties*, USDA, Economics, Statistics and Cooperative Services, Rural Development Report No. 10

Rowland, D.T. (1979), *Internal Migration in Australia* (Australian Bureau of Statistics, Canberra)

Rowley, G. (1971), 'Central Places in Rural Wales', *Annals, Association of American Geographers*, 537-50

Sadler, P.G. and Mackay, G.A. (eds.) (1977), *The Changing Fortunes of Marginal Regions* (Institute of Sparsely Populated Areas, Aberdeen)

Sarre, P. (1981), *Second Homes. A Case Study in Brecknock* (Open University, Milton Keynes)

Saville, J. (1957), *Rural Depopulation in England and Wales 1851-1951* (Routledge and Kegan Paul, London)

Shucksmith, D.M. (1981), *No Homes for Locals* (Gower, Farnborough)

Schwarzwellar, H.K. (1979), 'Migration and the Changing Rural Scene', *Rural Sociology*, 44, 7-23

Sewel, J. (ed.) (1979), *The Promise and the Reality – Large Scale Developments in Marginal Regions* (Institute for the Study of Sparsely Populated Areas, Aberdeen)

Shaw, J.M. (ed.) (1979), *Rural Deprivation and Planning* (Geobooks, Norwich)

Sherwood, K.B. (1982), *Housing and Population Turnover in a Rural Environment: a Study in Northamptonshire*, PhD in progress, University of Leicester

Standing Conference of Rural Community Councils (1978), *The Decline of Rural Services* (National Council of Social Service, London)

Thomas, J.G. (1972), 'Population Change and the Provision of Services' in Ashton, J. and Long, W.H. (eds.), 109-19

Thrift, N.J. (1979), 'Unemployment in the Inner City: Urban Problems or Structural Imperative? A Review of the British Experience' in Herbert, D.T. and Johnston, R.J. (eds.), *Geography and the Urban Environment*, vol. 2 (Wiley, London)

Todd, D. (1981), 'Rural Out-migration and Economic Standing in a Prairie Setting', *Institute of British Geographers*, 5, 446-65

Todd, D. (1981), 'Rural Out-migration in Southern Manitoba: a Simple Path Analysis of Push Factors', *Canadian Geographer*, 25, 252-66

Van Bemmel, A.B. (1980), 'A Welfare Geographical Approach of Living in Rural Areas – Some Remarks', paper presented at the Eleventh European Congress for Rural Sociology, Helsinki

Veldman, J. (1981), 'Space and Society in Rural Areas', Paper presented at the Eleventh European Congress for Rural Sociology, Helsinki

Vining, D. and Kontuly, T. (1978), 'Population Dispersal from Major Metropolitan Regions: an Interested Comparison', *International Regional Science Research Review*, 3, 49-73

Vining, D., Pallone, R. and Plane, D. (1981), 'Recent Migration Patterns in the Developed World', *Environment and Planning*, 13A, 243-50

Walker, A. (ed.) (1979), *Rural Poverty: Poverty, Deprivation and Planning in Rural Areas* (Child Poverty Action Group, London)

Walker, G. (1977), 'Social Networks and Territory in a Commuter Village, Bond Head, Ontario', *Canadian Geographer*, 21, 329-50

Walmsley, D.J. (1978), 'The Influence of Distance on Hospital Usage in Rural New South Wales', *Australian Journal of Social Issues*, 13, 72-81

Walmsley, D.J. and McPhail, I.R. (1976), *A Geography of Hospital Care in New South Wales*, New England Research Series in Applied Geography, No. 44

Webber, R. (1977), *Cumbria Social Area Analysis*, Technical Paper 22 (Planning Research Advisory Group, London)

Webber, R. and Craig, J. (1976), 'Which Local Authorities Are Alike?' *Population Trends*, 5, 13-19

Weekley, I.G. (1977), 'Lateral Interdependence as an Aspect of Rural Service Provision: a Northamptonshire Case Study', *East Midland Geographer*, 6, 361-74

Welsh Office (1964), *Depopulation in Mid-Wales* (HMSO, London)

Wenger, C. (1980), *Mid-Wales: Deprivation or Development: A Study of Patterns of Employment in Selected Communities* (University of Wales Social Science Monograph No. 5, Cardiff)

White, P.L. (1981), 'Migration at the Micro-scale: Intra-parochial Movements in Rural Normandy, 1946-54', *Transactions of the Institute of British Geographers*, N.S.6, 451-70

Wibberley, G.P. (1972), 'Rural Activities and Rural Settlements', mimeographed paper presented at the Town and Country Planning Association's Conference, 16-17 February, London

Wilkinson, K.P. (1978), 'Rural Community Change' in T.R. Ford (ed.), 115-25

Williams, J.D. and Sofranko, A.J. (1979), 'Motivation for the Immigration Component of Population Turnaround in Nonmetropolitan Areas', *Demography*, 16, 239-55

Williams, R., Bloor, M., Horobin, G. and Taylor, R. (1980), 'Remoteness and Disadvantage: Findings from a Survey of Access to Health Service in the Western Isles', *Scottish Journal of Sociology*, 4, 105-24

Williams, W.M. (1964), 'Changing Functions of the Community', *Sociologia Ruralis*, 4, 229-310

Winchester and Central Hampshire Community Health Council (1981), *Second Survey of Health Services in Rural Areas in the Central Hampshire Health District* (Winchester)

Zelinsky, W. (1962), 'Changes in the Geographic Patterns of Rural Population in the United States 1790-1960', *Geographical Review*, 52, 492-524

Zuiches, J. and Rieger, J.H. (1978), 'Size of Place Preferences and Life Cycle Migration: a Cohort Comparison', *Rural Sociology*, 43, 618-33

8 RECREATION

M.F. Tanner

Background

The origins of the recreation demands made on the surrounding rural areas by the modern metropolis can be traced back a century or more, yet the widespread provision of resources to meet these demands is largely a phenomenon of the second half of the twentieth century. It is true that the national park concept dates from at least 1864 when the Yosemite Valley was granted to the State of California for the purposes of preservation and public use, an intention that was subsequently incorporated into the Act establishing Yellowstone as the first true national park in 1872. Other national parks were created before the end of the century and by the time the National Park Service was established in 1916 the system included 15 national parks and 27 national monuments, mainly in the west and distant from the large eastern centres of population. Similar motives lay behind the creation of public forest lands in 1891, which became National Forests in 1905. But while public use was an integral part of the national park concept, the primary purpose of such designations was to protect areas of wonderful or spectacular scenery from destructive exploitation. Indeed it has been argued that the protection of such 'natural wonders' should be interpreted more in terms of their being a substitute for the cultural heritage of buildings and other historic monuments that the new nation lacked than as an act of preservation for its own sake (Runte, 1979). Only after the creation of the National Park Service were clear policies formulated for the management of such areas which included the provision of recreational facilities, while the US Forest Service did not begin to integrate recreation with other uses until after 1918. Parallel developments also took place in Canada where 28 national parks were established, mainly between 1885 and 1935, with the intention of preserving the best examples of the nation's heritage, although recreational use was also included as an objective from the start.

The North American idea of the national park spread rapidly in the period following the First World War, a process which accelerated still further after 1945. New categories of protected land were also added to the American and Canadian systems in the form of state and provincial parks, wildlife refuges and similar areas. By 1975 the *United Nations List of National Parks or Equivalent Reserves* included a total of 1,325 such areas in more than 100 countries (International Union for the Conservation of Nature, 1975). About one half of these were in the countries of the developed world, nearly half of which were in North America where the criteria adopted meant that more than 200 state and provincial parks were included.

Such figures tend to obscure the very different ways in which the national park concept has been interpreted in different parts of the world. Only in some newly developing countries of Africa, South America and parts of Asia, which were still going through the process of colonisation and settlement, was it possible to follow rigidly the North American practice of setting aside extensive areas of undeveloped and unpopulated land, the character of which would be protected by retention in public ownership. In areas where there was a long history of human occupance and most land was already in some form of productive use, different approaches had to be adopted. This applies particularly in Europe where only in some mountainous regions, like those of northern Norway and Sweden, was it possible to establish national parks which included extensive roadless areas which retained much of their natural character. Elsewhere national parks have usually been designated in upland areas which offer attractive landscapes coupled with marginal agriculture and declining rural populations.

Such an approach has been adopted in France and in England and Wales where there was an active amenity movement during the interwar period which pressed for the establishment of national parks on the North American pattern as a means of preserving both scenery and wildlife (Sheail, 1975), but the National Parks and Access to the Countryside Act, 1949 made provision for a very different kind of protected area. Following the recommendations of John Dower, national parks would be extensive areas of relatively wild country offering scenery of high quality in which most of the land would remain in private ownership and where agriculture, forestry and rural industries would continue to play a role in moulding the landscape (Dower, 1945). The essential character of such areas would be protected primarily by the application of strict planning controls and public land acquisition would be largely restricted to the provision of access. Only limited funds were to be made available for the development of recreational facilities and these would be administered by a new National Parks Commission, which was given a predominantly advisory role. During the 1950s 10 national parks of this kind were established covering some 8 per cent of the land area, all but one of which were in the uplands, while a further 8 per cent has been included in Areas of Outstanding Natural Beauty designated to protect other areas of high scenic quality, again with an emphasis on planning control. A similar pattern has emerged in Japan where the first National Parks Law was passed in 1931 under which 12 national parks were established by 1941, mainly in mountainous areas. The primary purpose of the Japanese national parks was to protect areas containing the nation's best scenery and by 1972 about 8 per cent of its land surface was covered by such designations. Here too both natural and cultural landscapes were included and the emphasis has been more on the control of development than on the provision of recreational opportunities. Other areas of valued scenic quality are protected by their inclusion in 'Quasi-National Parks' designated under a new Law of 1957 (Simmons, 1975).

Emergence of Problems

The effect of these developments was that by the 1950s the majority of industrialised countries possessed two kinds of land resource which offered opportunities for outdoor recreation. The first was the system of urban parks found in most of the larger cities of the industrialised nations which had been created mainly in the late nineteenth and early twentieth centuries to cater for the outdoor recreation needs of urban dwellers. The second was the system of national parks and similar areas where recreational provision was usually regarded as subsidiary to the protection of wildlife, natural features and scenic quality and which, by their nature, were often distant from the main centres of population.

In most countries this situation caused few problems for some time after the Second World War, for the lack of personal mobility meant that visitors to the national parks were relatively few in number and those that did visit them usually used public transport. Even in North America, where people began to travel in large numbers during the inter-war period and where tourist-oriented facilities were created at an early stage in some of the more popular national parks, serious problems did not really emerge until the early 1950s. These problems may be directly attributed to the social changes that accompanied the rise of the affluent society. Not only did people have higher incomes to spend on recreational activity but changes in the working pattern meant that greatly increased amounts of leisure time were available. This occurred both because of a reduction in the working week and because of the extension of annual paid holidays, but the most important factor was probably the widespread adoption of the two-day weekend, which meant that a significant block of leisure time was regularly available. The use of this leisure time was influenced most strikingly by the rising levels of private car ownership, for these gave an increasing proportion of the population a mobility they had previously lacked. Mobility was also increased by the construction of new high-speed roads, like the interstate highway system in the United States and the motorways of Western Europe, while the rapid expansion of air travel gave recreation and tourism an international dimension.

The result of these changes was not only a rapid rise of participation in recreational activities of all kinds, especially those which take place in the open air, but a significant change in the nature and location of these activities. People were no longer content to rely solely on the recreational opportunities provided by their local environment, but used their new-found mobility to travel to areas beyond the city, particularly at weekends. The great advantage of the motor car was its flexibility, for it offered freedom from the fixed routes and rigid timetables that characterise most forms of public transport. The weekend pleasure trip therefore emerged as a recreational activity in its own right and quickly became the dominant form of outdoor recreation in most industrialised countries. The availability of the private motor vehicle also had a dramatic effect on holiday patterns, for not only did it permit relatively cheap family travel over long distances but it freed participants from a dependence on traditional resorts, so that holidays were often focused on regions rather than single centres, with some becoming nomadic in character. Even in Japan, and more recently in some countries of Eastern Europe, where growing prosperity was not immediately accompanied by a rapid rise in private car ownership, the existence of efficient public transport systems meant that large numbers of urban dwellers could use their increased leisure time to make trips into the countryside.

The largely unforeseen impact of the growth in leisure time coupled with improvements in mobility was therefore greatly to increase recreational pressures on the countryside, particularly those areas that were regarded as being especially attractive or which had a clearly-defined recreational role. These pressures were felt most acutely in areas that were readily accessible from the main centres of population, but they also had serious effects on the more distant national parks and similar resources. Most obvious was the congestion and overcrowding which resulted from a concentration of high levels of use into limited areas at particular times of the year. Existing facilities were rarely designed to accommodate the large numbers of visitors that now sought to use them, a problem that applied both to routeways and to the parking areas, picnic sites, viewpoints and other nodal points to which most of these visitors were attracted. The motor car was also a source of noise and, especially where large-scale parking areas were provided, represented a significant visual intrusion into landscapes of high amenity value, a problem that was accentuated by its use as a means of transport for recreational equipment of all kinds. In North America, in particular, the growing use of motor homes, self-contained trailers, powerful motor-boats and specialised recreational vehicles also had an impact on the environment of such areas. In some cases, these pressures led directly to the physical deterioration of the resource, but more important was the encouragement which they gave to the construction of additional accommodation, access roads and other permanent facilities, especially in areas where tourism had become a significant element in the local economy. As a result, it became increasingly difficult to reconcile the provision of recreational opportunities with the primary objective of protecting valued resources from development and change. Rather different problems occurred in countries like Britain where there are few areas with a clearly-defined recreational role and most outdoor recreation takes place on land in multiple use. Here the buildup of recreational pressures sometimes led to serious conflicts with agriculture and other rural interests, while there was growing concern about the urbanisation of routeways and settlements in and around the national parks. Similar problems occurred in Japan where cultural features often acted as the main focal points in national parks (Simmons, 1975).

Outdoor Recreation Resources Review Commission

These problems emerged first in the United States and their origins can be traced back at least to the early 1920s when the attraction of visitors was adopted as an objective by the National Park Service and led to the promotion of tourism and development of permanent facilities. Many visitor facilities were provided in national parks and other areas under the public works programmes of the 1930s and the immediate response to the rapid growth of demand in the post war period was the provision of further facilities, often with little regard to the long-term effects on the environment of such areas (Runte, 1979). Only in the mid-1950s was this approach seriously questioned as a result of growing concern both that the nation's recreation resources were inadequate to meet the demands upon them and that some of the most valuable were being irreparably damaged by unsuitable development.

It was therefore suggested that a more comprehensive and wide-ranging approach was needed to the provision of outdoor recreation opportunities and in 1958 the Outdoor Recreation Resources Review Commission (ORRRC) was set up by Congress to investigate the problem. When the Commission reported back in 1962 it concluded that, while considerable areas of land were available for outdoor recreation, these did not effectively meet existing needs, either because of the way they were managed or because their concentration in the West made them inaccessible to much of the urban population. Without national action this situation was likely to get worse, for it was estimated that participation in outdoor recreation would experience a threefold increase by the end of the century, while additional facilities were urgently needed close to the large metropolitan areas. The Commission's recommendations therefore included the formulation of a nationwide outdoor recreation policy, the expansion, modification and intensification of present programmes, the introduction of a federal grants-in-aid programme and the establishment of a Bureau of Outdoor Recreation within the Department of the Interior to co-ordinate the national effort (ORRRC, 1962).

ORRRC presented the results of its work in the form of a summary report supplemented by 27 special studies and these led not only to a growth of public interest in outdoor recreation but also to changes in policy. They also probably did more than anything else to stimulate research in an area that had been largely neglected in academic circles. Their most immediate effect was on the methodology used in the study of outdoor recreation, for ORRRC had adopted a threefold approach involving the use of inventory surveys to determine the supply of resources, the use of questionnaire surveys to assess demand, and the examination of particular resource or policy questions that were regarded as being of special concern. This approach was widely copied in other countries, although nowhere has it been carried out so comprehensively. The ORRRC report therefore ushered in a first phase of outdoor recreation research with an emphasis on basic fact-finding exercises that lasted until the early 1970s.

The Study of Supply

ORRRC's examination of the availability of resources was based on an inventory of 24,000 non-urban publicly designated recreation areas, of which 5,000 of the largest were selected for further study to evaluate their present use and potential for development. Elsewhere national inventory surveys have sometimes been carried out, as in the Netherlands (Van Onzenoort, 1973), but generally a more piecemeal approach has been adopted. In Britain, for example, a great deal of resource information has been assembled by the Regional Sports Councils and other agencies, but its collection has been largely uncoordinated and most detailed studies have focused on the recreational use of particular categories of resource, like the coastline (Tanner, 1969), forests (Lloyd, 1972) and beaches in Scotland (Ritchie and Mather, 1976). Only in Canada has the evaluation of recreation resources been undertaken as part of a comprehensive national survey of land capability. Here the Canada Land Inventory was begun in 1963 as a co-operative federal-provincial programme with the broad objective of classifying the nation's land as to its potential for such uses as agriculture, wildlife, forestry and recreation (Burbridge, 1971).

Perhaps the most striking thing to emerge from such studies was the great diversity of the resources used for outdoor recreation, a diversity that is reflected in both their physical nature and in their ownership and land-use patterns. In the United States more than 80 per cent of the land used for recreation is publicly owned, although this ownership is shared among a great variety of agencies. This land includes both areas like national and state parks with a clearly-defined recreational function and areas in multiple use where recreation is generally subordinate to such primary objectives as grazing, timber production, water resource development and the protection of fish and wildlife. Outside North America the pattern is usually rather different, particularly because of the limited extent of public ownership and emphasis on multiple use. Nevertheless, in most industrialised countries a similar diversity of resources, both natural and man-made, is used for recreation, including parks, nature reserves, trails and footpaths, forests, rivers, lakes and reservoirs, beaches, historic buildings and areas of unenclosed mountains and uplands. As in the United States, the pressure of rapidly growing recreational demands has frequently caused problems, both because of existing management practices and because of spatial imbalance in the availability of recreational opportunities. This applies particularly in such countries as Britain and Japan where the designation of national parks had emphasised the protection of natural resource values, so that these tended to be located in upland areas away from the main centres of population.

The Study of Demand

ORRRC had based its conclusions about demand on two kinds of information. Most important were the data from a personal interview survey of 16,000 persons carried out in 1960. This use of comprehensive national surveys was subsequently copied in other countries. In Britain, for example, two such surveys were carried out in the mid-1960s (British Travel Association, 1967; Sillitoe, 1969), although with rather different objectives, and these were supplemented by a number of regional surveys, some of which provided detailed analyses of recreation patterns (eg, North West Sports Council, 1972). Such surveys not only indicated broad levels of participation in outdoor recreation, but revealed the great diversity of activities involved. They also provided information about personal background and economic status that was used to make forecasts which suggested that the rapid rates of growth characteristic of many activities would continue for the foreseeable future.

The second kind of information was that derived from surveys at individual sites. This approach had been used by ORRRC as part of its inventory of resources and such visitor surveys became the most common means of collecting information about participation in outdoor recreation, both in North America and elsewhere. Generally a questionnaire was used to collect data about age, sex, occupation, party size, origin of trip, mode of travel and frequency of participation with a view to both determining the characteristics of participants and providing information of use in planning and management. In some cases, studies were also made of the growth of particular types of activity, like water-based recreation (Tanner, 1973), in which information was brought together from a wide range of existing sources, an approach that has been referred to as 'data-dredging' (Coppock and Duffield, 1975).

Response to ORRRC and Similar Studies

The immediate effect of the ORRRC report and similar studies elsewhere was to stimulate public interest in outdoor recreation and a review of policies relating to the recreational use of rural resources. Such reviews commonly led to attempts to improve the planning machinery, to the allocation of increased public funds, to the provision of recreational opportunities and to the creation of new forms of facilities. In the United States the Bureau of Outdoor Recreation was set up in 1962 to carry out various planning functions, including the co-ordination of federal and state programmes and the preparation of a comprehensive nationwide outdoor recreation plan. Additional finance to assist federal and state agencies in recreational provision was provided for by the Land and Water Conservation Fund set up in 1964, while various new categories of protected land and recreation resources were created during the 1960s. These included the establishment of National Recreation Areas administered by the National Park Service, primarily with the intention of providing recreational opportunities in areas more accessible to the bulk of the urban population than the existing national parks, and the creation of the National Wilderness Preservation System in 1964, the National Trails System in 1968 and the National Wild and Scenic Rivers System, also in 1968. A similar response was reflected in Canada in the expansion of the provincial park system during the 1960s and in Japan by the creation of Quasi-National and Prefectural Parks under the 1957 Law (Simmons, 1975).

In Britain, what was in effect a review of recreation policies was published in 1966 in the form of a government White Paper on *Leisure in the Countryside* (Ministry of Land and Natural Resources, 1966), although a more comprehensive review of the whole sport and leisure field was carried out in the early 1970s by a Select Committee of the House of Lords (Select Committee, 1973). The 1966 White Paper proposed a considerable increase in central government expenditure on national parks and other outdoor recreation facilities, which would be channelled through a reconstituted National Parks Commission whose responsibilities would be extended to cover the whole of the countryside. It also proposed the creation of a new system of country parks which would be established and administered by local authorities but with grant aid from the central government. Such country parks would provide outdoor recreational opportunities in a rural setting and would be located so that they could act as 'honeypots' by absorbing high intensity recreational pressures that might otherwise fall on the national parks and other areas of vulnerable countryside. These proposals were implemented by the Countryside Act, 1968 which established the Countryside Commission for England and Wales, while a similar Act of the previous year created the Countryside Commission for Scotland. The growing importance of leisure activities was also reflected in the establishment of a national Sports Council in 1965 and subsequently in the creation of a system of Regional Sports Councils which attempted to co-ordinate the work of local authorities over the whole sport and recreation field, a role that was formally recognised in 1976 when they were renamed Regional Councils for Sport and Recreation and were given responsibility for preparing regional recreation strategies. Further Acts in the late 1960s and early 1970s reorganised local government and land-use planning, which indirectly affected the provision of recreational opportunities in the countryside, while the Forestry Commission and new Regional Water Authorities were given clearly-defined recreational functions.

Generally the assumption underlying such responses was that in the long term the problem caused by rapidly growing demand could only be solved by the provision of new recreational opportunities. This led to an early emphasis on demand-based planning in which research activity was focused particularly on the assessment and prediction of demand. The primary objective of planning was therefore often regarded as the matching of supply with the levels of demand revealed by surveys. Such an approach represented a marked departure from the resource-based planning that had characterised the evolution of national park systems in most countries.

Assessment and Prediction of Demand

Much early planning was based on the assumption that not only could existing demand be accurately measured but that this could be projected into the future with some degree of confidence. Various methods for the measurement of demand have been developed in both Britain and North America (Lavery, 1975), some of which have been subject to detailed examination at the local level (eg, Burton, 1971), but a number of problems remain. Generally, demand was expressed simply in terms of the total number of participants and only limited information was provided about the frequency of participation, which can be a variable of great significance. There were also problems related to the definition of activities, while the reliance on broad national surveys meant that there were sampling problems in recording minority pursuits, except where detailed local studies were carried out (eg, Duffield and Owen, 1971). More important is the fact that what was usually being measured was not 'demand' in the economist's sense of a quantity related to price, but merely existing use or participation.

These problems were accentuated when attempts were made to predict future levels of use. At the national level there was much reliance on trend extrapolation techniques which were usually based not on the projection of participation itself but on that of the socio-economic characteristics of the population that appeared to be related to participation in outdoor recreation. There was therefore considerable emphasis in surveys at both the national and site levels on the collection of data about such variables as age, sex, income, occupation, educational level and car ownership of which participation was thought to be a function. A number of forecasting techniques were developed using this method, but problems were encountered both in the measurement of some of these socio-economic variables and in the making of assumptions about the future relationship between them. The lack of adequate time-series data, especially about participation, was also a serious limiting factor. One possible solution to these problems was to try to identify groups of people with similar patterns of recreational behaviour, but it has become increasingly obvious that socio-economic factors alone do not adequately explain patterns of participation in outdoor recreation. For example, when cluster analysis was used to group people with similar socio-economic and recreational use characteristics with a view to estimating future demands, it was found that for the majority the most important factor in determining changes in recreational behaviour was a category labelled 'tastes and preferences' rather than such easily measured variables as age, income and education.

There was also considerable interest in the prediction of recreational use at the local level, especially where a new facility was proposed. In some cases, special studies of particular sites were carried out, but most progress has been made in connection with more general work on accessibility and the assessment of recreational benefits. The influence of mobility on recreation patterns, and especially the importance of the motor car, which in North America accounts for well over 90 per cent of all leisure travel, attracted the attention of research workers at an early stage. It has been shown that the car is an important influence on both pleasure trips and holiday patterns (Wall, 1972), but most work has been focused on its use for day visits to the countryside. An important concept here is that of the 'recreational hinterland', for it has been argued that the countryside should be regarded as an ecological extension of the city (Mercer, 1970). Most local surveys placed considerable emphasis on the origins of visitors to recreation sites and these data were used to examine travel patterns (eg, Colenutt, 1969). Various studies have demonstrated the existence of distance-decay relationships around large cities, as in Greer and Wall's (1979) examination of the destination of day-trippers, second-homers, campers and resort holidaymakers from Toronto, but it has also been shown that for many recreationists the journey itself is a valuable part of the recreation experience (eg, Cheshire and Stabler, 1976). Other workers have focused on the influence of new road construction on mobility, as in Britain where motorways have significantly increased the accessibility to the national parks for day-trippers (Jackson, 1970). Accessibility can also be important at the local level, for a study in Scotland showed that the most heavily used beaches were those easily visible from the road or near the main tourist centres (Ritchie and Mather, 1976).

Interest in the assessment of recreation benefits originated in the United States where it developed mainly in connection with multipurpose water resource projects, especially since the Federal Water Project Recreation Act, 1965 recognised recreation as a valid purpose of such schemes. Only rarely has attention turned to other kinds of resource, like McConnell's (1977) attempt to estimate benefit functions from beaches in Rhode Island. Benefits have most commonly been derived from the concept of 'willingness to pay', although this is itself difficult to identify because most outdoor recreation opportunities are available without charge. Occasionally, willingness to pay has been estimated by direct interview techniques, but the most common approach is to use travel costs as surrogate. Generally an assumption is made of a direct relationship between distance and cost, so that the further participants are prepared to travel the greater the benefits they derive. Data are collected about travel patterns which are used to produce a demand schedule, usually associated with the name of Clawson (Clawson and Knetsch, 1966), and this may be used both to assess current benefits and to predict changes in demand that might result from changes in cost. This technique has been used in a number of studies using empirical data, most commonly in North America but also in Britain where it has been suggested that the recreation benefits from a proposed new reservoir may represent a significant part of the social return (Anderson, 1975). There has been a longstanding debate about the appropriate methodology for such studies, but a number of serious problems remain. Various workers have questioned Clawson's assumption that a recreation journey can be regarded simply as a cost (eg, Cheshire and Stabler, 1976), while other problems concern the measurement of time and the complex relationship between supply and demand (Knetsch, 1974). Most studies have also encountered boundary problems in defining the area within which data are collected, for few recreation sites are sufficiently isolated for the influence of competing facilities to be ignored. The impact on other sites of both existing and new facilities is imperfectly understood and this has sometimes led to studies of individual sites being based on the assumption of a closed system.

Outdoor Recreation Research Since the Early 1970s

By the early 1970s the study of outdoor recreation had attracted the attention of research workers from a number of disciplines, including geography, economics, sociology, planning and ecology. Its methodology was still mainly that of the social sciences and there had been considerable investment in surveys of one kind or another which had provided a great deal of essentially descriptive information. The analysis of this information had certainly served to clarify the issues involved, but had not led to any major breakthrough in concepts or methodology. It has been suggested, for example, that a decade of research into the demand for recreation had done little more than identify areas of ignorance (Rodgers, 1975). There had also been much discussion of the relationship between recreation and leisure time, but more important had been the emergence of the 'recreation experience' as an integrating concept. The classic formulation of the outdoor recreation experience is that of Clawson and Knetsch (1966) who suggest that it has five distinct phases: (1) anticipation, (2) travel to the site, (3) on-site activity, (4) travel back, and (5) recollection. This description clearly relates primarily to the pleasure trip and has been criticised for its emphasis on travel, but is important for its recognition that outdoor recreation provides an experience that cannot be measured simply in terms of the physical activity which takes place at the site.

There was now a growing recognition that the factual nature of much outdoor recreation research meant that it was not necessarily leading to greater understanding, better prediction of future trends or more effective planning and management. The result was the adoption of a more balanced approach in which the emphasis was on a search for explanation rather than description (Coppock, 1980). More rigorous analytical techniques were used to explore both recreational behaviour and the nature of the resources on which it depends, while growing attention was paid to the complex interaction between them. Particular interest was also expressed in recreational satisfaction and in the basic motivational factors underlying leisure activity (Cherry, 1978).

This change of emphasis may also be related to changing economic circumstances, for outdoor recreation research had largely developed against a background of investment in the recreation resource base, which reflected rising prosperity and the general expectation of continuing rapid growth in participation. The deepening recession that affected most industrialised countries following the oil crisis of 1973 led to changing political attitudes towards public expenditure and a marked slowing down in the growth of outdoor recreation. As a result it appeared that the earlier optimistic forecasts of demand would not be realised, particularly because of the likely effect of rapidly increasing petrol prices on the leisure use of the private motor car upon which so much outdoor recreation depends. It therefore became clear that further investment in new facilities would be limited and that the provision of additional recreational opportunities would depend increasingly on the more effective management of existing resources.

Developments in Planning and Management

The need for management-oriented research also stemmed from changing attitudes towards the resources used for recreation. The US National Park Service had begun to review its park management policies in the mid-1960s, leading to a change of emphasis from the provision of visitor facilities to the application of ecological management techniques to safeguard wildlife and natural features. An emphasis on the maintenance of the natural environment was also an inherent requirement for the management of the new wilderness and wild and scenic rivers systems, which had expanded rapidly since their inauguration; by 1979 more than 19 million acres of land were included in 190 wilderness areas, while a total of 2,318 miles of wild and scenic rivers had been designated and others were under study. In order to achieve a balance between the interests of visitors and the maintenance of resource qualities, the National Park Service developed the use of master plans, usually based on a zoning system. This approach has been widely adopted in other areas. In Canada, for example, a five-class zoning system for land use in national parks was established as early as 1964 (Yapp and Barrow, 1979) and a land-use classification has formed the basis of master plans for New Zealand national parks (Millar, 1973). Zoning according to degrees of access has also been a feature of the emerging French national park system (Richez, 1973), while the management plans for British national parks were introduced by the Local Government Act, 1972 (Hookway, 1978). The preparation of such management plans, whether for areas where resource values are dominant or for intensive-use areas like the British country parks, requires a clear understanding of both the nature of the resource and the needs and behaviour of visitors.

A planning process for outdoor recreation has been most fully developed in the United States where ORRRC's 1962 report was adopted as the first 'Nationwide Outdoor Recreation Plan'. The second Nationwide Plan was prepared by the Bureau of Outdoor Recreation in 1973 (Bureau of Outdoor Recreation, 1973) and the third by the Bureau's successor, the Heritage Conservation and Recreation Service, in 1979 (Heritage Conservation and Recreation Service, 1979). Since 1965 each state has also produced a 'Statewide Outdoor Recreation Plan' covering a five-year period. In Britain, many local authorities prepared reports on outdoor recreation as part of the structure plan process, but there has been no national recreation plan although the government published a new review of policies as a White Paper on *Sport and Recreation* in 1975 (Department of the Environment, 1975). Only in Scotland has an attempt been made to produce a broad strategic plan for the development of facilities. Here the Countryside Commission for Scotland (1974) has proposed the creation of a park system comprising four main elements, traditional urban parks, 'countryside parks', 'regional parks' and 'special parks' similar to the national parks of England and Wales.

Such a planning approach can be important for identifying policy issues and determining national priorities, but the majority of decisions affecting the development of facilities are taken by the agencies responsible for management. The provision of outdoor recreation opportunities has therefore largely depended on a division of management responsibilities which is generally characterised by great complexity at both the national and local levels. In the United States all the big federal land management agencies, like the Bureau of Land Management, Forest Service, National Park Service and Fish and

Wildlife Service, have clear recreational responsibilities, while some 80 other agencies provide co-ordinating, financial, planning and other services related to outdoor recreation. Most states also provide outdoor recreation opportunities through their park and forest systems, while other federal agencies, such as the Bureau of Reclamation, US Army Corps of Engineers and Tennessee Valley Authority, are important at the regional level. It was this complexity that led to the establishment of the Bureau of Outdoor Recreation as a co-ordinating body, a role that was taken over by the Heritage Conservation and Recreation Service until its abolition in 1981.

In Europe the main land-managing agency is usually the state forest service and in some countries, such as the Netherlands (Simmons, 1975), this has played a leading role in recreational provision. The position is a little different in Britain where, although the Forestry Commission is the largest landowner, recreational opportunities are also provided by the British Waterways Board, Regional Water Authorities (Tanner, 1977), the Nature Conservancy Council and the quasi-public National Trust which own or control large areas of land. Local authorities are also very important, both through the direct provision of facilities and through their operation of the land-use planning system. No single co-ordinating agency has been established to deal with this overlapping of responsibilities. Instead there are a number of advisory bodies, like the two Countryside Commissions, Sports Council, Nature Conservancy Council, Water Space Amenity Commission and Tourist Boards, each concerned with a particular area of interest, although these too tend to overlap. There have been few studies of how this institutional system works, except within the context of countryside planning generally (Blacksell and Gilg, 1981), although some attention has been paid to the question of objectives. The Countryside Review Committee (1976), which examined rural planning in the mid-1970s, concluded that policies stemmed from the goals of individual agencies rather than from objectives related to the countryside as a whole. Both in Britain and the United States much of the outdoor recreation resource base is managed by multi-functional agencies whose objectives tend to be related more to their non-recreational activities than to the quality of the recreation experience or the requirements of users.

Continued Interest in Demand and Behaviour

The emphasis on large-scale planning exercises in the United States meant that there was continued interest in demand surveys. Between 1970 and 1977 at least 65 major surveys were carried out by federal, state and other agencies, but the lack of a consistent methodology prevented their use for the assessment of trends. An attempt to remedy this was made by the Heritage Conservation and Recreation Service in 1977 when it initiated a new Nationwide Outdoor Recreation Survey as part of the preparatory work for the third Nationwide Plan (Heritage Conservation and Recreation Service, 1979). There has been similar concern about the lack of time-series data in Britain, although the main result has been the inclusion of leisure questions in the *General Household Survey* in 1973 and 1977 (Birch, 1979), while the Countryside Commission carried out its own survey in 1977 (Countryside Commission, 1978).

Attention has also been focused on the factors which affect the incidence of demand at individual sites. In Britain work has continued on the origin, frequency, timing and destination of recreation trips and there has been particular interest in the impact of rising petrol prices during the 1970s. An examination of use at Forestry Commission sites found little evidence of major change (Collings and Grayson, 1977), but a study of fishing at a water supply reservoir showed it to be highly price-elastic (Shucksmith, 1979a). The long-term impact is not yet clear, but the available evidence suggests that there has been some decline in day and weekend trips to the more remote rural areas, while the use of the car for holiday travel has been little affected (Shucksmith, 1979b). A similar conclusion has been reached in the United States where it was found that recreational travel continued to increase during the 1970s, though at a slower rate. This suggested that the main effect of rising petrol prices would be seen in the recreational activity of those with lower incomes and in the decreased use of the more distant resource-based areas (Heritage Conservation and Recreation Service, 1979).

In North America the emphasis has been more on visitor behaviour, particularly in wilderness and other backcountry areas, for it has been argued that the management of such resources should be based on the analysis of behavioural systems (Peterson and Lime, 1979). This has led to growing interest in models that take account of recreational behaviour. The modelling of travel patterns of visitors to the countryside developed at an early stage (eg, Colenutt, 1969), but during the 1970s increasing use was made of simulation models for the management of wilderness areas designed to minimise both impact and encounters between visitors (Romesburg, 1974). More recently this approach has been adapted for use in the management of wild and scenic rivers (Carls, 1978). Simulation models have not yet been used in this way in Britain, although they may have some potential in the management of recreational footpath systems. Given this emphasis on behaviour, it is somewhat surprising that the nature of the recreation experience itself has received so little attention. Hautaluoma and Brown (1978) have tried to identify the components of the deer hunting experience, while Hammitt (1980) has produced results which support the original concept of outdoor recreation as a multi-phase experience.

The need to provide more effective public participation in the formulation of management policies for recreation areas has also stimulated research into the perceptions and attitudes of visitors. Questionnaire surveys have been used particularly to examine visitor attitudes towards management alternatives for wilderness areas and wild and scenic rivers. For example, a study of visitors to two California wilderness areas revealed considerable support for a permit system designed to limit use (Stankey, 1979).

The underlying theme of behaviour is similarly evident in the studies which have been made of conflicts between user groups. Trails and rivers appear to be particularly susceptible to such conflicts, especially where recreationists using motorised equipment come into contact with those following more traditional pursuits. This kind of conflict has become most acute in North America, mainly because of the growing use of off-road recreation vehicles (ORRVs), like snowmobiles, trail bikes and dune-buggies, but they have also occurred in Britain in the use of some ancient unmetalled routeways and navigable rivers. Detailed examination of these conflicts usually reveals that they are more complex than would appear at first sight. For example, a study of the conflict between snowmobilers and cross-country skiers in North America found that the problem stemmed not just from physical incompatibility but also from class differences and different attitudes towards the environment (Knopp and Tyger, 1973). Similarly a study of the conflict between coarse fishermen and boat users on the Norfolk Broads has shown the importance of social and psychological factors (Owens, 1978).

Research into the Management of Resources

The recognition that recreational use can have serious environmental consequences has led to growing interest in the resource base and its management. Particular attention has been paid to the nature of wilderness, mainly in North America but also in Australia (Smith, 1977). Wilderness management presents special problems, for it has been argued that, while the overriding goal must be to maintain the integrity of natural ecological processes over a fairly large area, it is also important to examine the attitudes, preferences and perceptions of visitors with a view to minimising their impact (Lucas, 1973). This approach leads logically to the introduction of rationing schemes, like the limitations on the number of entry permits to the Boundary Waters Canoe Area that were introduced in 1976 in order to prevent resource deterioration and to maintain the quality of the wilderness experience (Hulbert and Higgins, 1977).

Concern about the maintenance of the resource base has also led to research into impact, for outdoor recreation may affect the environment in which it takes place in various ways. Most obvious is the physical deterioration of the resource, the importance of which varies according to habitat type (Satchell, 1976), but a concentration of use can also lead to noise, traffic congestion and visual intrusion into the landscape. There has been particular interest in the effects of erosion and trampling on vulnerable ecosystems, like sand dunes (Trew, 1973) and the margins of water areas (Tivy, 1980), while in North America research has focused on the impact of visitor use on trails and campsites (eg, Foin *et al.*, 1977) and on the effects of ORRVs, like snowmobiles (Butler, 1974).

Closely associated with the concept of impact is that of recreational capacity, which has received a great deal of attention because of its obvious value to planners and managers. The underlying assumption of much early work was that capacity figures for recreation sites could be defined with some degree of precision and that the failure to do so owed more to the lack of reliable data than to any conceptual weakness. As the study of capacity developed, it became apparent that it is an extremely complex concept that is related to both the physical characteristics of the resource and to the social attitudes and perceptions of users. Burton (1974) has suggested that the recognition of separate landscape, ecological and perceptual capacities could aid the development of management policies, while Hardy and Penning-Rowsell (1974) examined recreational capacity as an operational concept and identified a need to experiment through the application of capacity standards in field situations. In the United States the problem was examined by the Bureau of Outdoor Recreation (1977) which concluded that most research had served more to identify areas of uncertainty than to provide definitions of carrying capacity that could be used by planners and managers. It therefore adopted a pragmatic approach which explicitly rejected the idea of a universal standard and instead offered practical guidelines for determining the 'optimum recreation carrying capacity' which was defined as 'the level of use most appropriate for both the protection of the resource and the satisfaction of the user'.

A rather different concern has emerged in Britain and some other European countries where there is statutory protection of national parks and similar areas based on essentially subjective decisions about landscape quality. An early attempt to develop a more objective approach to the assessment of scenery by Linton (1968), has been shown to have potential for development (Gilg, 1975). Various other techniques have been proposed, some of which have been compared in a study of south-east Devon (Blacksell and Gilg, 1975). More recently a computer-based information system has been used to delineate recreational landscapes in Scotland (Duffield and Coppock, 1975), while the landscape of part of Spain has been evaluated using grid techniques (Ramos *et al.*, 1976). Problems remain, for it has been shown that different individuals value scenic quality in different ways (eg, Preece, 1980) and it has been suggested that a distinction should be made between 'landscape character', which is a description of land forms and land use that can be quantified, and 'landscape quality' which is a purely subjective idea (Liddle, 1976). There has also been concern about changes in the landscape, especially those resulting from new agricultural practices which can dramatically alter the character of both national parks and other rural areas. Gilg (1979) has attempted the field mapping of landscape change in part of Devon, while Parry (1977) has examined the historical process of change at the moorland edge.

Conclusions

Since the early 1960s there has been a major expansion in the availability of outdoor recreation opportunities throughout the developed world, reflecting both the pressures of demand stimulated by growing prosperity and the increasing concern about the quality of the environment. The nature of this expansion has been much influenced by what has happened in North America, even though the high standard of living and public ownership of extensive areas of undeveloped land give outdoor recreation there a special character. There are a number of different types of outdoor recreation resource in most industrialised countries, although national parks have retained a strong symbolic value, whether they are intended to protect outstanding examples of wild country, like those of North America, or whether they mainly comprise areas of cultural landscape, like those of Britain and France. While the lack of reliable forecasting techniques means there is some uncertainty about the future of outdoor recreation, it is likely to remain an important use of rural land. At the same time, it appears that the designation of major new national parks is coming to an end in most such countries, partly because of changing political attitudes and the slackening in the growth of participation, but mainly because relatively few suitable areas remain that have not already been given national park or similar status. The emphasis is likely to be increasingly on the provision of intensive-use areas, like the British country parks or the American National Recreation Areas, designed to cater for the immediate recreational needs of a predominantly urban population. There is also growing recognition of the value of trails and footpaths in providing access to rural areas without heavy investment in land acquisition and management. Long-distance routes, like Britain's Pennine Way and America's Appalachian Trail, have a symbolic value similar to that of the national parks, but attention is likely to be focused increasingly on the provision of 'recreational' trails and footpaths at the local level, especially where these can be linked to other facilities.

The provision and management of these resources is dominated by public agencies in every country. During the early 1970s it was anticipated that there would be a growing role in the provision of intensive-use facilities for the private sector, but its achievements have been largely limited to essentially urban-oriented developments, like theme parks, and to the provision of accommodation and other ancillary services. Much outdoor recreation, especially in Europe, takes place on land in multiple use and this has led to a growing awareness that the maintenance of cultural landscapes suitable for recreation depends on effective management, but some of the most interesting management initiatives have been a response to the special problems of American wilderness areas. The development of effective systems of multi-purpose management has been hindered by an inadequate understanding of both recreational behaviour and the ecological functioning of outdoor recreation resources. Much recent research has been designed to remedy this deficiency and, after the early dominance of geographers, the study of outdoor recreation has become increasingly multi-disciplinary. This has led to the application of new concepts and research methodologies, although it has been suggested that there is still room for a more intellectually rigorous approach (Patmore and Collins, 1980). It has also been noted that, in spite of heavy investment in management-related research, the application of results is rare, partly because managers and researchers do not always agree on which information is useful (Schreyer, 1980). Certainly two decades of research have increased our understanding of outdoor recreation, but the Heritage Conservation and Recreation Service (1979) noted in the *Third Nationwide Outdoor Recreation Plan* that recreation remains a complex concept and that there was 'little agreement on what recreation means or should mean'.

References

Anderson, R.W. (1975), 'Estimating the Benefits from Large Inland Reservoirs' in *Recreational Economics and Policy*, ed. G.A.C. Searle (Longman, London)

Birch, F. (1979), 'Leisure Patterns 1973 and 1977', *Population Trends*, 17, 2–8

Blacksell, M. and Gilg, A.W. (1975), 'Landscape Evaluation in Practice: the Case of South-east Devon', *Transactions of the Institute of British Geographers*, 66, 135–40

Blacksell, M. and Gilg, A.W. (1981), *The Countryside: Planning and Change* (George Allen and Unwin, London)

British Travel Association-University of Keele (1967), *Pilot National Recreation Survey, Report No. 1* (University of Keele, Staffordshire)

Burbridge, V. (1971), 'Methods of Evaluating Rural Resources: the Canadian Experience', *Journal Town Planning Institute*, 57, 257–9

Bureau of Outdoor Recreation, US Department of the Interior (1973), *Outdoor Recreation: a Legacy for America* (US Government Printing Office, Washington)

Bureau of Outdoor Recreation, US Department of the Interior (1977), *Guidelines for Understanding and Determining Optimum Recreation Carrying Capacity* (US Government Printing Office, Washington)

Burton, R.J.C. (1974), *The Recreational Carrying Capacity of the Countryside*, Occasional Publication 11 (University of Keele, Staffordshire)

Burton, T.L. (1971), *Experiments in Recreation Research* (George Allen and Unwin, London)

Butler, R.W. (1974), 'How to Control 1,000,000 Snowmobiles', *Can. Geog. J.*, 88, 4–13

Carls, E.G. (1978), 'A Simulation Model of Wild River Use', *Leisure Sciences*, 1, 209–18

Cherry, G.E. (1978), 'Aspects of Recreation Research' in *Recreational Planning and Management in the New Local Authorities*, ed. A.J. Veal (Centre for Urban and Regional Studies, University of Birmingham)

Cheshire, P.C. and Stabler, M.J. (1976), 'Joint Consumption Benefits in Recreational Site "Surplus": an Empirical Estimate', *Regional Studies*, 10, 343–51

Clawson, M. and Knetsch, J.L. (1966), *Economics of Outdoor Recreation* (Johns Hopkins Press, Baltimore)

Colenutt, R.J. (1969), 'Modelling Travel Patterns of Day Visitors to the Countryside', *Area*, 2, 43–7

Colenutt, R.J. and Sidaway, R.M. (1973), *Forest of Dean Day Visitor Survey*, Forestry Commission Bulletin 46 (HMSO, London)

Collings, P.S. and Grayson, A.J. (1977), *Monitoring Day Visitor Use of Recreational Areas*, Forestry Commission (HMSO, London)

Coppock, J.T. (1980), 'The Geography of Leisure and Recreation' in *Geography Yesterday and Tomorrow*, ed. E.H. Brown (Oxford University Press)

Coppock, J.T. and Duffield, B.S. (1975), *Recreation in the Countryside: a Spatial Analysis* (Macmillan, London)

Countryside Commission (1978), *Digest of Countryside Recreation Statistics 1978* (CCP 86, Cheltenham)

Countryside Commission for Scotland (1974), *A Park System for Scotland* (Battleby)

Countryside Review Committee (1976), *The Countryside, Problems and Policies: a Discussion Paper*, Department of the Environment (HMSO, London)

Department of the Environment (1975), *Sport and Recreation*, Cmnd 6200 (HMSO, London)

Dower, J. (1945), *National Parks in England and Wales: Report by John Dower*, Ministry of Town and Country Planning, Cmd. 6628 (HMSO, London)

Duffield, B.S. and Coppock, J.T. (1975), 'The Delineation of Recreational Landscapes: the Role of a Computer-based Information System', *Transactions of the Institute of British Geographers*, 66, 141–8

Duffield, B.S. and Owen, M.L. (1971), *Leisure + Countryside = A Geographical Appraisal of Countryside Recreation in the Edinburgh Area* (Department of Geography, University of Edinburgh)

Foin, T.G. *et al.* (1977), 'Quantitative Studies of Visitor Impacts on Environments of Yosemite National Park, California, and Their Implications for Park Management Policy', *Journal of Environmental Management*, 5, 1–22

Gilg, A.W. (1975), 'The Objectivity of Linton Type Methods of Assessing Scenery as a Natural Resource', *Regional Studies*, 9, 181–92

Gilg, A.W. (ed.) (1979), *Policies for Landscapes under Pressure* (Northgate Press, Bury St Edmunds)

Greer, T. and Wall, G. (1979), 'Recreational Hinterlands: a Theoretical and Empirical Analysis' in *Recreational Land Use in Southern Ontario*, ed. G. Wall, Department of Geography, University of Waterloo, Publications Series 14, 227–46

Gum, R.L. and Martin, W.E. (1977), 'Structure of Demand for Outdoor Recreation', *Land Economics*, 53, 43–55

Hammitt, W.E. (1980), 'Outdoor Recreation; Is It a Multi-phase Experience?', *Journal Leisure Research*, 12, 107–15

Hardy, D. and Penning-Rowsell, E.C. (1974), 'Recreation Capacity as an Operational Concept', *Cambria*, 1, 17–27

Hautaluoma, J. and Brown, P.J. (1978), 'Attributes of the Deer Hunting Experience: a Cluster-analytic Study', *Journal Leisure Research*, 10, 271–87

Heritage Conservation and Recreation Service, US Department of the Interior (1979), *The Third Nationwide Outdoor Recreation Plan: The Assessment* (US Government Printing Office, Washington)

Hookway, R.J.S. (1978), 'National Park Plans: a Milestone in the Development of Planning', *The Planner*, 64, 20–2

Hulbert, J.H. and Higgins, J.F. (1977), 'BWCA Visitor Distribution System', *Journal of Forestry*, 75, 338–40

International Union for Conservation of Nature (1975), *1975 United Nations List of National Parks and Equivalent Reserves*, IUCN Publications New Series No. 33 (Morges, Suisse)

Jackson, R. (1970), 'Motorways and National Parks in Britain', *Area*, 4, 26–9

Knetsch, J.L. (1974), *Outdoor Recreation and Water Resources Planning*, American Geophysical Union, Water Resources Monograph 3

Knopp, T.B. and Tyger, J.D. (1973), 'A Study of Conflict in Recreational Land Use: Snowmobiling vs Ski-touring', *Journal Leisure Research*, 5, 6–17

Lavery, P. (1975), 'The Demand for Recreation: a Review of Studies', *Town Planning Review*, 46, 185–200

Law, S. (1967), 'Planning for Outdoor Recreation in the Countryside', *Journal Town Planning Institute*, 53, 383–6

Liddle, M.J. (1976), 'An Approach to Objective Collection and Analysis of Data for Comparison of Landscape Character', *Regional Studies*, 10, 173–82

Linton, D.L. (1968), 'The Assessment of Scenery as a Natural Resource', *Scottish Geographical Magazine*, 84, 219–38

Lloyd, R.J. (1972), 'The Demand for Forest Recreation' in *Lowland Forestry and Wildlife Conservation*, ed. R.C. Steele, Monks Wood Experimental Station, Symposium No. 6, The Nature Conservancy, London, 93–108

Lucas, R.C. (1973), 'Wilderness: a Management Framework', *Journal Soil and Water Construction*, 28, 150–4

McConnell, K.E. (1977), 'Congestion and Willingness to Pay: a Study of Beach Use', *Land Economics*, 53, 185–95

Mercer, D.C. (1970), 'Urban Recreational Hinterlands: a Review and Example', *Professional Geographer*, 22, 74–8

Millar, D.D. (1973), 'Environmental Planning and Management Policies', *Town Planning Quarterly*, 33, 12–19

Ministry of Land and Natural Resources (1966), *Leisure in the Countryside: England and Wales*, Cmnd. 2928 (HMSO, London)

North West Sports Council (1972), *Leisure in the North West* (Deansgate Press, Salford)

Outdoor Recreation Resources Review Commission (1962), *Outdoor Recreation for America*, 27 volumes (US Government Printing Office, Washington)

Owens, P.L. (1978), 'Conflict Between Norfolk Broads Coarse Anglers and Boat Users: a Managerial Issue' in *Social Issues in Rural Norfolk*, ed. M.J. Moseley (University of East Anglia, Norwich)

Parry, M.L. (1977), *A Framework for Land Use Planning in Moorland Areas*, Department of Geography, University of Birmingham, Occasional Publications, 4

Patmore, J.A. (1970), *Land and Leisure in England and Wales* (David and Charles, Newton Abbot)

Patmore, J.A. and Collins, M.F. (1980), 'Recreation and Leisure', *Progress in Human Geography*, 4, 91–7

Peterson, G.L. and Lime, D.W. (1979), 'People and their Behaviour: a Challenge for Recreation Management', *Journal of Forestry*, 77, 343–7

Preece, R.A. (1980), 'An Eavluation by the General Public of Scenic Quality in the Cotswolds A.O.N.B.: a Basis for Monitoring Future Change', Department of Town Planning, W.P. 48, Oxford Polytechnic

Ramos, A. *et al.* (1976), 'Visual Landscape Evaluation: a Grid Technique', *Landscape Planning*, 3, 67–88

Richez, G. (1973), 'Les parcs natural dans le Sud-Est de la France', *Méditerannée*, 12, 119–35

Ritchie, W. and Mather, A.S. (1976), 'The Recreational Use of Beach Complexes of the Highlands and Islands: a Note', *Scottish Geographical Magazine*, 92, 61–3

Rodgers, H.B. (1975), 'The Demand for Outdoor and Active Recreation: a Decade of Research in Retrospect' in *Environment, Man and Economic Change*, ed. A.D.M. Phillips and B.J. Turton (Longmans, London), 479–91

Romesburg, H.C. (1974), 'Scheduling Models for Water Recreation', *Journal of Environmental Management*, 2, 159–78

Runte, A. (1979), *National Parks: the American Experience* (University of Nebraska Press, Lincoln)

Satchell, G.E. (1976), *The Effects of Recreation on the Ecology of Natural Landscapes*, Nature and Environment Series 11 (Council of Europe, Strasbourg)

Schreyer, R. (1980), 'Survey Research Recreation Management – Pitfalls and Potentials', *Journal of Forestry*, 78 (6), 338–40

Select Committee of the House of Lords on Sport and Leisure (1973), *Second Report* (HMSO, London)

Sheail, J. (1975), 'The Concept of National Parks in Great Britain 1900–1950', *Transactions of the Institute of British Geographers*, 66, 41–56

Shucksmith, D.M. (1979a), 'The Demand for Angling at Derwent Reservoir, 1970 to 1976', *Journal Agricultural Economics*, 30, 25–37

Shucksmith, D.M. (1979b), 'Petrol Prices and Rural Recreation in the '80's', *National Westminster Bank Quarterly Review*, February, 52–9

Sillitoe, K.K. (1969), *Planning for Leisure*, Government Social Survey (HMSO, London)

Simmons, I.G. (1975), *Rural Recreation in the Industrial World* (Edward Arnold, London)

Smith, P.E. (1977), 'A Value Analysis of Wilderness', *Search*, 8, 311–17

Stankey, G.H. (1979), 'Use Rationing in Two California Wildernesses', *Journal of Forestry*, 77, 347–9

Tanner, M.F. (1969), 'Coastal Recreation in England and Wales' in *Coastal Recreation and Holidays*, Countryside Commission (HMSO, London), 1–78

Tanner, M.F. (1973), *Water Resources and Recreation* (The Sports Council, London)

Tanner, M.F. (1977), *The Recreational Use of Water Supply Reservoirs in England and Wales* (Water Space Amenity Commission, London)

Tivy, J. (1980), *The Effect of Recreation on Freshwater Lochs and Reservoirs in Scotland* (Countryside Commission for Scotland, Battleby)

Trew, M. (1973), 'The Effects and Management of Trampling on Coastal Sand-dunes', *Journal Environmental Planning and Pollution Control*, 1, 38–49

Van Onzenoort, A.A.H.C. (1973), 'Outdoor Recreation Planning in the Netherlands', *Planning and Development in the Netherlands*, 7, 50–63

Wall, G. (1972), 'Socio-economic Variations in Pleasure Trip Patterns: the Case of Hull Car-owners', *Trans. Inst. Br. Geographers*, 57, 45–57

Yapp, G.A. and Barrow, G.C. (1979), 'Zonation and Carrying Capacity Estimates in Canadian Park Planning', *Biological Conservation*, 15, 191–206

9 RESOURCE EVALUATION AND MANAGEMENT

P.J. Cloke

Rural Resources and Management Agencies

Whenever the words *rural* and *resource* are juxtaposed, ensuing discussion is liable to be bogged down with uncertainties over semantic definition. For example, many commentators will argue that 'there is no unambiguous way of defining rural areas' (Moseley *et al.*, 1977) even though it is commonly recognised that 'the distinction between "urban" and "rural" areas is deeply rooted in the psychology of most attempts at any regional or local subdivision' (Shaw, 1979). Similarly, no one definition of a resource has remained generally acceptable through time, largely because resources are culturally defined and abstract in concept. For the purposes of this chapter these issues are side-stepped and a rather narrow view of rural resources is adopted. Physical elements in the countryside (water, minerals, land, and so on) have traditionally been subject to conventional approaches to resource management (Mitchell, 1979; O'Riordan, 1971; and Simmons, 1974). Only more recently has access to social and economic opportunities in the countryside been recognised within the rural resource framework (Whitby and Willis, 1978) and yet it is to this 'human' sector that planning and management processes have become increasingly attracted in post-war Europe and America. Although there appears to be a strong case for the *collective* appreciation of 'physical' and 'human' rural resources (Cloke and Park, 1980, 1983) this chapter adopts the rather underdeveloped notion of rural resources as those social and economic life-style opportunities where differential access by various income and age groups can lead to those problems collectively termed 'deprivation'.

The concept of rural deprivation has recently received a glut of detailed empirical attention (Runciman, 1972; Shaw, 1979; Walker, 1978; and Knox and Cottam, 1981) while discussions as to the exact meaning of the term have been further pursued by Moseley (1980a) and McLaughlin (1981). What is important to note here is that the rationalisation processes underlying deprivation have very often been directly and causally linked with the management of various rural

198

opportunities (or resources) in both the private and public sectors. Private-sector decision-makers have followed market trends which dictate that ever-increasing threshold populations are required for the profitable provision of services and facilities (for example, Harman's (1978) analysis of local village shops). In the public sector, resource rationalisation and the high per capita costs of servicing a scattered and small-scale rural population have again induced an outcome of deprivation in some cases, with the village school, the doctor's surgery and the local post office being withdrawn in favour of concentrations of junior and secondary education, three- or four-doctor health centres and more scale-efficient post offices in the larger towns. Moseley (1980) views decision-makers in the various agencies concerned with these outcomes as the *producers* of rural deprivation and argues that much more attention should be directed towards the institutions which allocate scarce resources and the macro-context of society's economic and political structure; that is, the factors which may be seen as actually dictating rural life chances.

With this background of rural resource managers whose decisions can often result in distributional outcomes, the study of resource management and evaluation in rural areas has become a crucial interface between geographical and planning studies (Cloke, 1979a; Harrison, 1977). Of particular relevance is the extent to which society's co-ordinating mechanism for overseeing resource distribution — the planning system — has been able both to derive socially and politically acceptable framework policies for rural resource allocation, and to *implement* those policies successfully. An appreciation of the scale of both of these tasks becomes apparent when it is realised that most of the important locational decisions concerning the basic resources which affect rural life-styles are taken *outside* the formal planning process by agencies other than planning departments and committees. The role of these agencies has been underemphasised in many of the assessments of rural planning performance, even though what actually happens in rural areas is largely a function of decisions made in this sphere of operation.

Moseley's (1979) discussion of agency decisions affecting accessibility emphasises several general factors which attribute importance to informal planning organisations:

1. No single planning agency is alone responsible for, or capable of, maintaining the welfare, well-being and general livelihood of rural communities.

2. For most relevant agencies, decisions affecting rural areas are usually secondary considerations, if they are considerations at all: other objectives are usually paramount or better defined.
3. In consequence of 1. and 2., decision-making tends to be fragmented and ill-focused on rural problems.
4. While there is no shortage of literature on the powers and policies of individual agencies, how they interact (or fail to interact) one with another has received little attention.
5. While in the medium or long term a more integrated and purposeful process of rural planning and management might be developed, in the short run policy measures cannot be more comprehensive or ambitious than the present institutional context allows.

Thus the tasks performed by resource allocation agencies are crucial to any consideration of the evaluation and management of socio-economic resources in rural areas.

This chapter offers insufficient scope for a detailed breakdown of rural decision-making agencies (Cloke, 1982). Some indication of the plethora of public-sector agencies to be considered is given in Figure 9.1, and the functions and workings of such organisations are well reviewed by the Association of County Councils (1979), the Association of District Councils (1978) and Woollett (1981). Perhaps less frequently acknowledged is the role of private-sector decision-making in rural areas. In fact, public- and private-sector activities are often interlinked. For example, public-sector attempts to generate new employment in rural areas are usually dependent on individual entrepreneurs or private companies to respond to the incentives involved. In turn, private-sector financial institutions will be required to underwrite the activity of the entrepreneur even if he is responding to public-sector incentives. An even more symbiotic private/public sector relationship occurs between the post office and village shopkeepers. Whether in harness with public agencies, or working independently the decisions made by private-sector individuals or organisations are of far-reaching importance to rural people. The decisions of firms of all sizes and key individuals (notably farmers) influence the availability of employment (Hodge and Whitby, 1981). Accessibility provision in rural areas is heavily reliant on independent bus operators who, by using family labour and multiple role employment methods generally enjoy lower overheads than their National Bus Company equivalents (Nutley, 1981). The housing sector, too, is greatly influenced by the private sector. Local authority planners can devise strategic policies for

Figure 9.1: Central and Local Government Responsibilities in Rural Areas

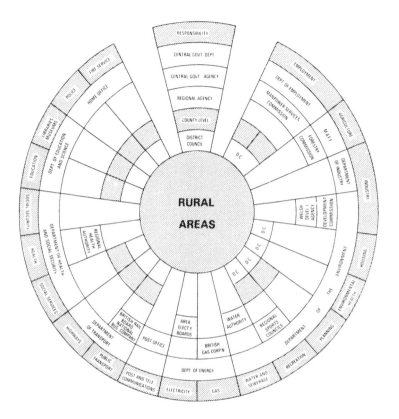

residential development, but the decision to develop lies elsewhere. District housing committees allocate council houses, but private developers supply the applications which planners permit or refuse. In this instance, it is private-sector landowning and development agencies (and their financial backers) who decide on the number, type and purpose of residential proposals (Winter, 1980). The importance of private-sector decision-making in rural areas in many ways undermines the ability of local authority planners to maintain effective control of resource distribution. Indeed, the agencies which, in reality, have the power to control the allocation of finance and other resources in rural settlements are those over which the formal planning process has least control. It is against this background of decision-making powers that attempts to evaluate and manage rural life-style opportunities should be measured.

Resource Evaluation

Economic Evaluation Methods

Attempts by government and local authority planners to influence the broad pattern of resource distribution in rural areas have largely been undertaken through the mechanism of first development planning and second structure planning at the county council level (see Chapter 10). Analysis of structure plan resource allocation policies (Blacksell and Gilg, 1981; Cloke, 1979; Gilder and McLaughlin, 1978; Parsons, 1979; and Rawson, 1981) clearly demonstrates that local authorities have been aware of the anti-concentration arguments presented to them during the 1970s but have been slow to realign their policies towards more dispersed distributions of rural resources. Evidently, the evaluation of alternative resource policies has in general failed to convince policy-makers that resource dispersal schemes are economically viable. Such evaluations have been predominantly economic in nature during the development plan and structure plan eras and is likely to continue in that vein given the current financial restrictions imposed on local authorities by central government. It is thus apparent that economic analysis of resource allocation policies in rural areas will continue to be of crucial importance to the future of rural communities.

This major role which should be played by economic analysis has not been matched by research effort from geographers and planners. What little research has been carried out has been directed towards empirical analyses of rural public service costs, as it is the allocation of these particular resources which, when allied to similar decisions over housing and employment, tend to govern the underlying spatial pattern of rural settlement policy. Curry (1981) discusses other possible research avenues:

A number of economic approaches to testing the key settlement hypothesis may be pursued. These include cost-benefit analysis, planning balance sheets, threshold analysis, linear programming and so on, but all of these approaches, and the pursuit of the often 'higher' objectives that their use entails (for example that of Pareto Optimality in the case of cost-benefit analysis) have commonly been a source of confusion with public authorities because of their degree of sophistication and the way in which their results might be interpreted.

In fact three main examples of the study of public service costs serve to illustrate the conceptual complexities and pitfalls which are endemic in this element of decision-making in rural planning.

The South Atcham Study. Warford's (1969) study of the South Atcham area of Shropshire attempted to predict the likely costs and benefits of various settlement and resettlement options. In particular he was concerned with the measurement of the effect of dispersed and rationalised settlement patterns on the provision costs of water supply, sewage disposal, education facilities, housing and transport services. In a rather raw cost-benefit analysis, he concluded that a relocation option which involved a resource concentration strategy would present the preferable scheme for a new water-supply system. Warford was careful to enunciate the limitations of his technique, which neverthe-less represents a landmark in the economic analysis of rural settlement strategies. This rather all-embracing method of analysis has, however, been subject to critical comment (Whitby and Willis, 1978; Willis, 1980). For instance, the assumption that relocation investment would be immediate tends to emphasise capital costs over budgetary costs and this has led to various ambiguities in the final evaluation of options. The South Atcham study also highlights the problems involved in trans-lating a raw cost-benefit analysis into useful policy indicators. Various policy options were considered, ranging from retention of the status quo to the relocation of population at various levels. When these options were ranked in terms of net social benefit, they presented a near-perfect inverse correlation with the ranking of options by decreas-ing budgetary cost. The degree to which a 'best' option can be isolated within this process is therefore almost totally dependent upon the selective criteria adopted by decision-makers. So, for a least-cost approach, decision-makers would have to accept sub-optimum social benefits, the levels of which can only be improved by higher cost inputs. The cost-benefit technique is an efficient method of ordering these variables, but is of little assistance in the selection of policy options and can therefore contribute little direct evidence to the evaluation of the debate between concentration or dispersal of resources in rural areas.

The North Walsham Study. An approach to resource evaluation based on public cost optimisation was pioneered by Norfolk County Council (1976) in their study of the North Walsham area. Instead of assessing the cost and benefits of relocating a *fixed* population, the study predicted the likely population *growth* for this area and undertook an evaluation of the relative costs and benefits associated with the accommodation of this population within different types of settlement pattern. The North Walsham study also represented an improvement over previous techniques in that policy-making guidelines were explicitly adopted prior to the commencement of option evaluation. Thus, 'the strategy which achieved the highest return on public spending by producing the most effective provision of services in relation to their cost was judged to be the "optimum" strategy' (Norfolk County Council, 1976). The results of this study were analysed by Shaw (1976) who outlined four principal resource allocation options:

1. *Concentration* – concentrating growth in the largest settlements.
2. *Dispersal 1* – a standard rate of growth in all villages.
3. *Dispersal 2* – making use of spare capacity in services, taking account of physical constraints.
4. *Dispersal 3* – developing villages of 500–800 population.

Capital and revenue costs of various public services and facilities were analysed for each of these four options. In addition it was admitted that other social costs and benefits did not lend themselves to economic cost analysis and so these were assessed in non-monetary terms. As might be expected from this initial division of variable measurements, the study's conclusions were also split into economic and social outcomes. In economic terms, the analysis suggested that the concentration option was less costly than the dispersal options so far as both capital and revenue costs were concerned. In social terms, strong arguments were found in favour of accepting the higher economic costs of resource dispersal. The difficulties of attaching comparative priorities to social and economic costs served to cloud both the policy impact of the study and its general value as a widely applicable evaluation technique.

In fact, the North Walsham study's economic cost conclusions have themselves attracted criticism. Gilder (1979), for example, notes that in assessing the future costs of servicing the various resource allocation options, the study omits those costs which would be incurred in maintaining services for the *present* population which is already tied to existing locations in the study area. In this way, capital investment for additional services may be overemphasised to the detriment of a realistic assessment of the future revenue costs generated by existing services. Gilder also criticises the study for assuming the continuation both of existing levels of service provision, and of the current balance of public and private sector contributions towards the costs of those services. Both of these assumptions may be proved invalid if rural settlement planning assumes a more flexible approach to resource allocation in the future. In view of these deficiencies, the apparent evidence offered by the North Walsham study in favour of resource concentration policies is of reduced impact in the general context of resource evaluation methodology.

The Bury St Edmund's Study. The most important and thought-provoking analysis in this series is Gilder's (1979) appraisal of resource allocation options in the Bury St Edmunds area of Suffolk. He made a very detailed empirical study of two areas of public service costs:

(i)   an investigation of public service costs in the area covering a single year;
(ii) an analysis of public service costs under different population strategies (including those of constant or declining population).

Education, sewage disposal, transport and community health services were selected for analysis, and an important element of the study was a cross-sectional analysis of the cost of these services in relation to settlement size. In effect a 'full costing' approach is used (in contrast to the relocation of present population in South Atcham and the additional costs of future growth in North Walsham).

Gilder's conclusions are in sharp contrast to those from South Atcham and North Walsham, and tend to lend support to his previous settlement policy analysis (Gilder and McLaughlin, 1978):

> the accommodation of future growth will be less costly if that growth is dispersed widely throughout the area, than if it is concentrated in Bury St. Edmunds and the larger villages. The marginal costs of making better use of existing schools and other fixed assets is likely to outweigh the benefits of concentrated development policies, even if economies of scale within services exist. The relative costs of maintaining services at existing levels for present populations far exceed the costs of accommodating new growth (Gilder, 1979).

As in all analyses of this kind, the exact methodology used to achieve these conclusions has been the subject of discussive criticism (Cloke and Woodward, 1981). In particular the exact nature of the measurement of costs and economies of scale has been scrutinised in view of the fact that the assessment of public sector budgetary costs can often diverge widely from the broader notion of 'economic' cost which may, perhaps, be more accurately represented by the use of opportunity costs. Furthermore, by ignoring the *exact* incidence of budgetary costs, analyses of economies of scale in rural resource allocation can exclude the consideration of distributional effects, which lie at the very heart of the themes of deprivation and disadvantage discussed earlier.

Gilder (1981) has legitimately defended the use of public sector costs in the Bury St Edmunds study on the grounds that 'Public authorities are, regrettably, most influenced in any decision by the costs that will actually fall on the public purse'. Similarly, he rejects the measurement of economic cost: 'The introduction of economic notions of cost would inevitably require a series of assumptions to be made about the distribution of costs and benefits. Questioning each of these assumptions would have provided a convenient refuge for all those who cling to the conventional wisdom'. This is a very telling comment as it suggests that particular modes of analysis for the evaluation of resource allocation alternatives appear to be better suited to the protagonists of one or other side of the concentration-dispersal debate, and that the selection of one such mode can therefore influence or at least provide scope for the shaping of the outcome. In fact this suggestion may be more apparent than real and does not detract from the fundamental importance of the Bury St Edmunds study which is an important milestone in the analysis of rural strategies. If further usage of the full costing technique confirms the study's pro-dispersal conclusions then the economic basis of key settlement policies will have been severely weakened, and the need to consider alternative rural resource allocation strategies will be stronger than ever. Clearly, however, the technical ambiguities which have clouded some of the central issues in the Bury St Edmunds study would necessarily do so in other areas without recourse to continual improvements in resource evaluation methods. There is still some progress to be made before an acceptable method of economic evaluation of rural resource allocation can be presented to decision makers.

Resource Evaluation in Planning

The three studies reviewed above appear to offer the prospect that scientific investigation is able to produce an objectively derived optimum pattern for the distribution of resources in rural areas. This rather academic view is effectively dashed when the political realities of central and local government decision-making are considered. Jenkins (1978) highlights this split between academic study and the realities of resource management processes:

> a positive advantage of evaluation would be to depoliticise a situation, to provide a cold rational appraisal of policy alternatives or policies *per se* outside the steam heat of emotion and ideology. Sadly, this is very much a false hope, a product of technocracy and scientism pushed forward by those who hanker after a managerial outlook and who fail to appreciate that there is really no such thing as an apolitical arena.

It is evident therefore than planning authorities are forced to take additional political factors into account in their resource allocation decisions. Most authorities readily admit this procedure and use it as a legitimisation for their failure to indulge in complex and detailed resource evaluation techniques. Hertfordshire County Council (1976), for example, describe four alternative strategies as possible options for their structure plan, but reject a methodological evaluation of these alternatives. Rather it is recognised that:

> the real decision about the basis for this Plan hinged upon reaching a compromise on various features of all four strategies. This compromise has to reflect a position of public and political acceptability, and a willingness by the many implementary agencies to operate within it.

Given the political nature of evaluation in the policy-making process, it would be naïve to expect planning authorities to divulge the exact details of how a preferred strategy was actually arrived at. Indeed, the compromises and trade-offs between resources, implementation and acceptability are often of an intangible nature which is difficult to measure or document. Most counties, however, have followed a three stage process of evaluation.

(i) The Alternatives. Almost without exception, county structure plans describe various alternative policies as the first stage of their policy evaluation process. The manner in which these alternatives are described is obviously very important in the arrival at a preferred option, and the widespread variation shown in the description of different structure plan alternatives demonstrates how perceptions of potential policy frameworks can vary from authority to authority. Hertfordshire County Council (1976), for example, established four *area-specific* alternatives ranging from the achievement of a balance between employment, housing and population growth by controls in both urban and rural areas, to an option of high population and employment growth in North and South-West Hertfordshire. In contrast, East Berkshire County Council (1980) chose to evaluate policies of no growth, limited growth, controlled growth and high growth *without* specific reference to the zoning of these growth levels within the structure plan area. The Cambridgeshire (1980) plan is different again. Here three *functional* strategies, centring on conservation, economic potential, and a reduction of social and spatial irregularities in living standards, are evaluated in comparison with existing policies.

Two important points are worthy of note at this 'alternatives' stage. First, very few plans outline alternative strategies using the concepts of resource concentration and dispersal which have formed the knub of more academically based evaluation procedures. Second, even when these concepts are adopted, the variation of potential rural framework policies is often overshadowed by the search for countywide development proposals. In fact, rural issues *per se* are given low priority at this stage, and needless to say, the omission of rural details at the beginning of the policy evaluation process has important ramifications for the way in which specific rural policies are subsequently generated to fit in with the selected countywide policy.

(ii) Comparison of the Likely Consequences. Assessing the likely consequences of alternative strategies would appear to be a stage where use could easily be made of academic studies which present evidence of the outcome of various policy mechanisms in rural areas. Any such potential is not visibly taken up in county structure plans, where analyses of the likely consequences of policy alternatives often constitute informed speculation of a low-key nature. Cornwall County Council's (1976) analysis of policy choices demonstrates the typical achievements of this stage of the evaluation process. Table 9.1

Table 9.1: Predicted Consequences of Policy Alternatives in the Cornwall Structure Plan

---

Choice A: To promote maximum economic growth and efficiency
8,000–17,000 new manufacturing jobs, 1975–91
18,000–38,000 new job opportunities
1,400–4,200 annual in-migrants
N.B. environmental considerations would have less priority.

Choice B: To remedy deficiencies and achieve economic stability
4,500–12,000 new manufacturing jobs, 1971–91
11,000–33,000 new job opportunities
800–3,600 annual in-migrants
N.B. development would be restricted to a limited number of settlements as a result of the need to concentrate the use of resources.

Choice C: To maintain the physical character of the county
'The more stringent environmental controls on industry might tend to inhibit growth in employment opportunities, leading to a lower rate of net migration by the economically active'.
N.B. environmental considerations would lead to a greater level of resource concentration, meaning a more economic provision of a wider range of services.

Source: Cornwall County Council, 1976.

shows the predicted consequences of three alternative strategies for the county, and suggests that in attempting to simplify these outcomes for the benefit of public participation, the inherent complexity and inaccuracy of growth predictions can be reduced to a range of achievements of various planning objectives. Any comparison of consequences which have been defined in this manner is therefore inevitably based on the political priorities attributable to different *objectives* rather than to different *outcomes*. It is thus the third stage of the evaluation process where resource allocation decisions are actually made.

(iii) The Achievement of Planning Objectives. Government advice (in the form of Structure Plans Note 8/72 from the Department of the Environment) suggests that a matrix model should be used to compare how alternative policies perform in relation to various planning objectives. A good example of how this advice has structured the policy evaluation methods used by planning authorities is presented by the Northamptonshire plan (1977). The evaluation was based on a quantitative analysis of how individual alternatives performed against certain

Table 9.2: Technical Evaluation of the Northamptonshire Structure Plan Alternatives

Objective	Performance measure	Order of achievement 1	2	3
Social objectives:				
1. Channel resources to disadvantaged areas and social groups.	Population outside New and Expanding Towns 1991.	Social	Environmental	Economic
2. Encourage the provision of a range of employment locations and opportunities throughout the County.	Improvement of Job/Worker Ratio 1971–91 (Average per Policy Areas).	Social	Environmental	Economic
3. Maintain the viability of existing commercial and employment centres.	(i) Growth in population in centres not benefiting from planned expansion 1975–91. (ii) Growth in jobs outside New and Expanding Towns 1971–91. (iii) Net outward commuting journeys 1991.	Social	Environmental	Economic
4. Ensure that every household has good access to employment locations.	Increase in commuting trips between Policy Areas 1971–91.	Social	Environmental	Economic
5. Ensure that every household has good access to shopping, educational, recreational and other community facilities.	Population growth in New and Expanding Towns, A.6 Towns and Key Centres.	Social	Environmental	Economic
6. Ensure a choice of residential types and locations.	Population growth in rural areas in relation to all growth outside New and Expanding Towns.	Social	Environmental	Economic
Economic objectives:				
7. Make the most efficient use of available resources.	Population growth Northampton and Corby 1975–91.	Economic	Environmental	Social
8. Ensure the efficient generation of new industrial and commercial development.	Job growth in New and Expanding Towns 1971–91.	Economic	Environmental	Social
9. Minimise the impact of development on agriculture.	(i) Residential land-take in New and Expanding Towns. A.6 Towns and villages classified as Key Centres and Moderate Growth villages in the Alternatives. (ii) Residential land-take in 'Remainder of Villages' 1975–91. (iii) Land-take for industrial use 1975–91. (iv) Land-take for all other purposes 1975–91.	Economic	Environmental	Social
10. Prevent the sterilisation of workable mineral deposits.	Mineral areas safeguarded in all Alternatives.	All alternatives perform equally well		
Environmental objectives:				
11. Protect and enhance the existing character of the villages and the rural environment.	Residential land-take in New and Expanding Towns. A.6 Towns and villages classified as Key Centres and Moderate Growth villages in the Alternatives weighted by landscape quality 1975–91.	Environmental	Economic	Social
12. Minimise the loss of high quality landscape.	Population growth in Key Centres and Moderate Growth villages 1975–91 within areas for Landscape Protection and Conservation.	Environmental	Economic	Social
13. Improve the quality of the urban environment.	No measure.	No measure		

Source: Northamptonshire County Council, 1977.

prescribed objectives. Three generic groups of alternatives (representing social, economic and environmental themes) were used in the analysis and the results of the exercise are summarised in Table 9.2. Inevitably each generic alternative, performed well when measured against its 'parent' set of objectives (economic policies achieve economic objectives and so on) although it is significant to note that the social alternative performed least well overall.

The usefulness of this kind of technique is openly questioned in the Northamptonshire plan: 'the differences in the level of achievement of objectives between alternatives was only marginal, since in relation to the overall level of growth up to 650,000, differences in population and job distributions between the alternatives were relatively small'. The utility of the evaluation exercise proved to be marginal since Northamptonshire's preferred strategy was most strongly influenced by the 'economic' alternative, not apparently as a result of the technical evaluation, but because it was the *safest option*:

Insomuch as the Economic Alternative had a pattern of development which was an extension of use of the existing infrastructure, and used established agencies to promote employment growth and concentrated population growth in areas where knowledge and control were relatively comprehensive, this Alternative handled uncertainty best.

The three stages of policy evaluation in structure plans demonstrate just how unsuccessful the academic concentration-dispersal debate has been in selling its 'advances', particularly in evaluation methodology, to county planning authorities. The evaluation methods adopted by structure plan teams suggest that previous policy precedents are by far the strongest determining factor in the preparation of new strategies for rural resource allocation. With one or two notable exceptions, the innovative and socially-oriented resource-dispersal strategies have been rejected in favour of more easily predictable and manageable extensions of current policy directions. This trend strengthens the inevitability that local-scale rural resource management will increasingly be determined through the implementation procedures of resource allocating agencies rather than as an integral part of a county strategy which is determined by wider structural issues.

Implementation and Rural Resource Management

An analysis of resource concentration strategies within both the development plan and structure plan eras suggests that the poor performance of key settlement policies may be at least partially explained in terms of problems of policy implementation. It is therefore important to note that equitable resource allocation *per se* policies (however evaluated) will not be sufficient to tackle the problems of rural deprivation and disadvantage. In many cases, poor implementation procedures can nullify beneficial resource distribution strategies. As Glasson (1979) notes: 'The test of the value of any plan, policy or strategy is the extent to which it influences the making and taking of decisions.' Although the study of implementation procedures by urban policy analysts has been of considerable importance for a number of years (Jenkins, 1978; Dunsire, 1978; Levitt, 1980), rural researchers have been slow to follow suit. It is only recently that rural commentators have fully acknowledged that written policy statements are not necessarily translated exactly by the planning decisions made at ground level, and that a rather murky divide exists between policy formulation and policy implementation. Detailed discussion of implementation processes are beyond the scope of this chapter, but the brief summary offered here can be extended by reference to the excellent set of essays edited by Barrett and Fudge (1981).

Two main implementation themes are particularly relevant to the understanding of rural resource management. First, it has become increasingly clear that policy implementation assumes different characteristics according to the type of planning which is pursued in rural areas. Minay (1979a) stresses that:

> there is, in fact, a conceptual difficulty about defining implementation without first determining one's theoretical stance: the idea of policy-making implies an identifiable outcome which represents implementation, but as one moves away from this concept towards one of continuous planning, implementation and policy making become less easy to distinguish until ultimately they merge.

He proceeds to identify four views of planning each of which encompasses different notions of implementation.

(i) Planning as a Response to Private Action involves a form of implementation in which the principal mode of action is that of controlling

the initiatives of private and public sector developers. Here, resources outside the decision-making powers of planners may be restricted or permitted *where they arise*, but are much more difficult to channel into locations suggested by overall resource strategies.

(ii) Planning as the Positive Promotion of Environmental Change is based on some dissatisfaction with development offered by the private sector and indicates a wish to bring about certain planning objectives through direct action. The importance of this form of planning is stressed by the Working Party on Rural Settlement Policies (1979): 'Positive rural strategies have been much less successful than development control policies, yet in many ways the implementation of positive policies is the most crucial element in rural settlement planning.' The lack of success in positive rural planning thus far suggests that if environmental change is to be promoted then implementation will involve the mobilisation of necessary resources for use in rural areas. Although this form of implementation has been evident, for example, in the inner city and new town programmes, there is little sign of similar resource allocations having been made in rural areas, and so the aims of 'positive planning' have tended to be pursued through the medium of co-operation with agencies where resources for rural improvement are available.

(iii) Planning as Co-ordination is particularly relevant in rural areas, where a multitude of agencies have the ability to distribute various resources and where the powers and resources available to planners are generally insufficient for direct action of a positive nature. Given this situation, it is the degree to which planners can successfully co-ordinate the actions of other agencies which has formed the yardstick against which implementation is measured. As each agency has its own aims, strategies and implementation procedures, the major task for planning is to achieve consistency and compatability between the various policy strands. This can be done either by the encouragement and persuasion of certain agencies to reconstitute their policies and actions into a more conforming pattern, or by a more radical reorganisation of the powers and resources vested in these various agencies into a more unified and 'controllable' structure (Lefebvre, 1978; Olsson, 1978).

(iv) Planning as Resource Management, which is most central to the theme of this chapter, is the most ambitious concept of the four, in

that it suggests that planning might act as a management system for physical and social resources. Minay (1979) argues that as yet this role for planning has been suppressed deliberately:

> One reason why governments opt for a relatively low level of co-ordination between agencies is no doubt the problem of resource management: the more co-ordination, the more arguments over resources. Organisations which are reasonably autonomous can get ahead and implement whatever seems important to them with the resources they possess without arguments that the resources would be better spent on something else.

If planning really were to assume the role of resource management, implementation would represent the process by which resources are allocated. This situation is wholly unrealistic in the current rural context. A large proportion of the resources which underlie rural life-styles belong to private sector agencies ranging from local to multi-national scales. These resources, therefore, currently evade public sector control. Even within the public sector, the competition for resources is a national one in which rural areas are often given low priority. As such it is not a reallocation of resources within rural areas, but a reallocation within society which is required to meet the needs of deprived sectors of rural communities.

These four different views of planning and implementation high-light the resource management achieved thus far in rural areas. The main emphasis has undoubtedly been on the functions of control and co-ordination of which the first has been more successfully achieved than the second. Moreover, the Working Party on Rural Settlement Policies (1979) appear to support a continuing priority for co-ordinative implementation in the future:

> Successful policy implementation clearly depends on an acceptance of a policy and a desire to make it work on the part of the public, local authority, statutory undertakers and central government, together with the right tools for the job. The means include efficient communication between the various bodies and organisations involved in rural development to ensure a common approach to problems, and effective financial management to ensure that the availability of funds is flexible enough to cover factory buildings or services when suitable opportunities present themselves.

The use of terms such as 'effective financial management' and a unified 'desire to make it work', however, suggest that planning as a positive promotion of change and ultimately as an agent for equitable resource allocation may not be redundant concepts in the future of rural communities.

The second broad implementation theme which is of particular relevance to rural resource management is that of focal problem areas where implementation difficulties have continually led to a breakdown in resource allocation procedures. Three such foci are readily identifiable in the rural context. First, the relationship between policy and implementation is a very close one, with rural resource allocation strategies being of little use if they are incapable of being implemented. Pressman and Wildavsky (1973) highlight this issue:

> The great problem as we understand it, is to make the difficulties of implementation a part of the initial formulation of policy. Implementation must not be conceived as a process that takes place after, and independent of, the design of policy. Means and ends can be brought into somewhat closer correspondence only by making each partially dependent on the other.

In rural Britain, policies of resource concentration have been adopted largely for pragmatic reasons, yet the means of implementing such policies, so as to achieve growth in the selected areas *and* to ensure that the new centralised opportunities were available to outlying rural communities were not an integral part of the policy itself. Similarly, strategic policies of resource dispersal would be hamstrung if no direct implementation procedures were made available.

The second focal problem of implementation concerns the level of powers and resources which are made available to policy-makers. It is evident that traditional resource concentration policies require both powers and resources which are currently outside the scope of local planning authorities. If demand for development exists, then the influence of planners is strong; if not, there is relatively little that can be done to induce it. Local authorities have some considerable power over the services which they provide (the problem here is more one of co-ordinating different departments and local government tiers) but have no direct control over water, electricity, post and telecommunications, and most health services. Similarly, they can do little about the closure or initiation of private-sector services, where no entrepreneurial demand exists. Planners are thus handicapped by deficiencies of power and resources. Positive intervention through subsidy, socio-economic experimentation and direct action has thus been given little emphasis during a harsh financial climate for local authorities. Moreover, there is some evidence to suggest that local authority involvement in positive socio-economic planning is not wholeheartedly welcomed by central government. Despite the central backing for experiments such as the RUTEX schemes in the transport sector, there is a continual fear that a public agency which acts beyond its powers may be punished by law and that an authority which spends beyond its means will be subjected to a claw-back operation through the withdrawal of rate support grants.

A third common problem in the implementation process is that of ensuring that decisions made by resource allocation agencies achieve some degree of consistency. Healey's (1979) analysis of the relationship between county council structure plans and district council local plans highlights one of the major conflicts which exists in this context.

> In practice . . . there is considerable tension between the structure plan and the local plan, district planning authorities often stretching the discretion available to them as far as they can. Development control decisions may subsequently undermine the policies of both plans. In other words, local plan and development control decisions can distort and override policies defined in a strategic framework, thus redefining policies.

Similar difficulties are encountered in the relationships between planning authorities and other agencies such as water authorities and education departments. These organisations have their own policy objectives, their own assessments of 'need' and 'priority' and their own long-term investment programmes. It is not surprising that conflict continually occurs when these specialist decision patterns are required to correspond to the spatial whims of planners, who are often viewed as just one other local authority department. Inconsistent decision-making has been rife in the development plan era of rural planning with the co-ordinated growth or conservation of particular settlements often being broken down by the actions of one dissenting resource agency. Furthermore, co-ordination is more easily achieved with an agreement to do nothing (a form of negative planning) than to collaborate with some positive socio-economic scheme. The strengthened co-ordinating role for planning encouraged by the structure plan system is designed to overcome these conflicts, but the widespread dissipation of decision-making powers creates extreme difficulty in assembling a comprehensive approach to the welfare of rural communities. Even if the scattered spread of rural resources could be effectively harnessed, the question remains as to whether sufficient resources exist, either within rural areas or in the proportion of national resources at presently allocated for that purpose to enable a positive response to rural problems. The final section takes up this theme.

Rural Resources and the Future

The post-war period in Britain has been characterised by a growing *awareness* of rural social and economic problems rather than a propensity to take *action* against them. This apparent lack of action is related to a general reluctance to acknowledge the political implications of rural disadvantage (Walker, 1978), and the lack of vision regarding the overall objectives which underlie the various attempts to maintain the presence of rural communities in Britain.

Given the task of alleviating the problems of the rural disadvantaged in the short term, and preparing a new socio-cultural framework for rural areas in the long term, rural resource managers have two broad avenues of approach, of which the first is to make use of existing resources.

Harnessing Existing Resources

How can new opportunities be provided on an equitable basis for rural residents given current resource bases? A detailed account of possible modes of action within this approach is presented elsewhere (Cloke, 1982), but three groups of possibilities may be summarised here.

New Initiatives from Existing Agencies. It can be argued that with a reorientation of priorities, the decisions taken by resource allocation agencies could result in additional benefits to disadvantaged rural groups. Planners themselves can assume greater degrees of advocacy, for example, in promoting particular forms of new housing through informal discussions with private developers. Indeed, a more flexible form of development control might be used to permit socially beneficial schemes such as Smigielski's (1978) self-supporting co-operative village. These changes in the attitudes and actions of formal planners will be thwarted by the internal survival mechanisms of local authority politicians unless a breakdown of convention and establishmentism occurs. A similar breakdown is required if other local authority resource agencies (dealing with housing, education, transport and the like) are to be persuaded to reallocate in favour of disadvantaged individuals and groups. For example, many local authorities have stockpiles of land under their ownership which could be utilised, at a cost to reserves but not to revenue, for local need housing schemes (as has been done in Tavistock, Devon). Local authorities can also do much by reorganising personnel so that willing individuals can be switched to the task of localised community co-ordination so that the valuable work carried out by Rural Community Councils (McLaughlin, 1979) can be continued and dramatically extended.

For public agencies outside the control of local authorities, the need is for an acceptance of flexible servicing schemes in rural areas, such as one-unit disposal schemes and peripatetic rather than centralised health-care services. In the private sector, the profit motive is not easily geared towards resource reallocation in rural areas, and subsidy or even nationalisation may be required before current private sector opportunities are significantly improved in all but fast-growing areas.

Inter-agency Experiments. Another usage of existing resources to reallocate in favour of disadvantaged groups is through positive planning experiments backed by various resource agencies. In many cases, such experiments have been based on the idea of placing a form of 'community worker' in rural areas. In the Cleator Moor area of Cumberland, for example, a mobile 'information and action' van has been established (Butcher, Cole and Glen, 1975) which offers various peripatetic information and representation services to the rural hinterland. The potential of the community work approach has been closely argued in the Cumbria case (Voluntary Action Cumbria, 1974) and has also found favour elsewhere (Hereford and Worcester Rural Community Project, 1978, 1980, 1981).

Self-help and Community Approaches. The report of the Wolfenden Committee (1978) on the future of voluntary organisations has highlighted the potential for rural self-help schemes. Although the political and moral arguments apparent in the shifting of resource responsibilities from government to community are complex and should not be underplayed (Dungate, 1980; Wilmers, 1981) there does appear to be a role for voluntary effort in the redistribution of rural resources. Fudge (1981) provides a useful summary of the situation:

mutual and community oriented alternatives, involving collective work and responsibility should be advocated and fought for not because they are cheaper (because they may well not be!) or because they relieve pressure on the statutory services (because pressure is needed for maintenance and improvement, and many will always rely on them) but because they are more sensitive services and they are often areas where innovation can occur. The doubt remains however that many groups only respond to symptoms rather than to underlying problems and that because of this the raising of consciousness among members leading to political analysis and action is unlikely to be forthcoming.

Self-help has flourished in rural areas. Woollett (1981) has provided a comprehensive summary of community services and facilities provided by voluntary effort in rural areas. Retail, transport, education, health care and community-care services have all been the subject of schemes, at least in experimental conditions. Shankland (1981) and Pearce and Hopwood (1981) have analysed the potential for community-based job creation schemes, and self-help housing experiments have been described by Winter (1980) and Kingham (1981). It has further been argued that self-help transcends the mere provision of opportunities, rather forming a type of social glue (Russell, 1975) for rural communities. However, the generalised limitations of approaches which attempt to harness existing resources apply equally to voluntary services, and as Shaw (1979) recognises: 'The solution to rural problems will remain partly a matter of resource allocation, particularly at national level.'

Attracting New Resources

The themes of resource allocation discussed above are all limited by the obligation that they operate within existing resource levels. However, it seems increasingly clear that deprivation and disadvantage in rural areas can only be effectively tackled either by attracting substantial additional resources to the treatment of problematic outcomes of socio-economic and political structures, or by changing the nature of these structures so that more equitable opportunity decisions become the norm; or indeed by both these measures together. Resources attracted to rural problems inevitably are withdrawn from elsewhere and so a realignment of both theory and practice is involved here. Potential realignments might be foreseen under three headings (Cloke, 1982).

Rural-urban Resource Allocation. One method of gaining additional resources is to plead a 'special need' and underprivilege in rural areas. The government's Countryside Review Committee (1977) could find 'no hard evidence that the countryside has been starved of resources in the past', whereas bodies such as the Association of District Councils (1978) claim clear evidence of rural resource starvation, of imbalanced rate support distributions and of urban bias in specific grant aid to deprived areas. This strong division of opinion has clouded the issue of whether rural areas deserve additional resources. A further complication to this situation is that little is known about the nationwide distribution of public expenditure and so the question of whether rural areas get their 'fair share' cannot be founded on an accurate information base (Moseley, 1980).

One way around the semantic and information problems involved here is to gain acceptance of the need to provide essential services universally. Thus, what is required is not additional *rural* resources, but positive intervention to benefit *all* disadvantaged groups be they urban, rural, peri-urban or belonging to any other special category which is fashionable. In this way no 'special' rural case is needed, because the rural disadvantaged should be given equal benefit from a national movement towards the redress of opportunity deficiency. This movement could be funded from local rates (Green, 1980) but it seems preferable that additional resources be derived from national taxation mechanisms which are more closely related to personal income. Any such positive equalisation of opportunities obviously requires fundamental changes in political opinion and, just as obviously, decision-makers will attempt to restrict this process to the minimum levels required by political acceptability and self-preservation. We are presented with the stark option of either continuing social disadvantage or precipitating some level of resource redistribution at a national and international scale.

Direct Action. Even if some redistributive impetus is achieved, the question remains of how the direct intervention of national resources is best used in the treatment of the causes as well as the symptoms of disadvantage in rural areas. One mechanism of resource use is through direct action by local authorities in a situation which the Royal Town Planning Institute (1979) regards as having enormous scope for informal and local initiatives. The ideological pros and cons of direct action have been given detailed discussion by Morley (1981), Byrne (1981) and Coopman (1981), but scope does seem to exist for direct action to provide the so-called 'necessary opportunities' in rural areas. Little research exists on this topic but the list should perhaps include nationalised, subsidised or co-operative employment initiatives; local authority direct-build schemes, or subsidy of co-operative groups, self-build groups or housing associations in the housing sector; provision of a package of informal transport services as a social service; co-ordination of a weekly market of peripatetic services in small settlements (Moseley, 1979); and subsidy or nationalisation of small shops (Kirby *et al.*, 1981) so that these too can be partially viewed as a social service.

Corporate Management. Although many of the policies described above are several steps removed from current conventional thinking, there have been recent moves towards a corporate approach to the management of rural areas. This would involve a more comprehensive form of decision-making and some form of corporate resource allocation in rural areas. The Working Party on Rural Settlement Policies (1979) lends its support to this reform:

> Costs would then be examined in terms of benefits and trade-offs between the various services to provide an overall financial assessment for a given policy package and, as necessary, a comprehensive analysis of its likely effects on individual groups or areas. The final stage would relate the overall policy package back to the individual sectors for any necessary policy adjustments to be made. Such an approach will not be achieved easily, since it will require a major change of emphasis in financing on the part of a number of public bodies. Nevertheless we consider it essential if the present problems and future needs of rural areas are to be dealt with satisfactorily.

This viewpoint is significant because it marks the small beginnings of radical thinking from a quasi-official body connected with rural resource allocation. Corporate rather than disparate resource management in rural areas could become a powerful method for the fulfilment of needs, and thus an equitable means of isolating and evaluating the priority needs of any rural community is a vital prerequisite for this new approach.

It is far easier to pinpoint the deficiencies of rural resource re-allocation than to offer potential improvements, particularly if these are to be seen as 'realistic' by decision-makers. Many of the alternatives mentioned above represent radical changes, yet it might be argued that radical changes are necessary because of the progressive trends of imbalanced resource distribution and social inequality in rural communities.

Each of the options discussed requires political decisions to be made, along with a clear recognition that the needs of each community may be different, and that the notion of a 'standard' solution to rural resource problems is meaningless in view of these differences. Moreover, the longer this decision is avoided, the more likely it is that rural areas will eventually become havens for the affluent minority rather than sources of wider opportunity.

References

Association of County Councils (1979), *Rural Deprivation* (ACC, London)

Association of District Councils (1978), *Rural Recovery: Strategy for Survival* (ADC, London)

Barrett, S. and Fudge, P. (1981), *Policy and Action: Essays on the Implementation of Public Policy* (Methuen, London)

Blacksell, M. and Tiley, A. (1981), *The Countryside: Planning and Change* (Allen and Unwin, London)

Butcher, H., Cole, I. and Glen, A. (1975), 'Information and Action Services for Rural Areas', University of York, Papers in Community Studies, No. 5

Byrne, S. (1981), 'Comment on Morley (1981)', *Town Planning Review*, 52, 306-7

Cambridgeshire County Council (1980), *County Structure Plan: Approved Written Statement*

Cloke, P.J. (1979), *Key Settlements in Rural Areas* (Methuen, London)

Cloke, P.J. (1979a), 'New Emphases for Applied Rural Geography', *Progress in Human Geography*, 4, 181-217

Cloke, P.J. (1980), 'The Key Settlement Approach: the Theoretical Argument', *The Planner*, 66, 98-9

Cloke, P.J. (1982), *An Introduction to Rural Settlement Planning* (Methuen, London)

Cloke, P.J. and Park, C.C. (1980), 'Deprivation, Resources and Planning: Some Implications for Applied Rural Geography', *Geoforum*, 11, 57-61

Cloke, P.J. and Park, C.C. (1983), *Resource Management in the Countryside: A Geographical Perspective* (Croom Helm, London)

Cloke, P.J. and Shaw, D.P. (1983), 'Rural Settlement Policies in Structure Plans' (forthcoming)

Cloke, P.J. and Woodward, N. (1981), 'Methodological Problems in the Economic Evaluation of Rural Settlement Planning' in Curry, N. (ed.), 1981

Coopman, S. (1981), 'Comment on Morley (1981)', *Town Planning Review*, 52, 308-9

Cornwall County Council (1976), *County Structure Plan: The Policy Choices*

Countryside Review Committee (1977), *Rural Communities* (HMSO, London)

Curry, N. (ed.) (1981), 'Rural Settlement Policy and Economics', *Gloucestershire*

Papers in Local and Rural Planning, no. 12, July
Curry, N. and West, C. (1981), 'Internal Economics of Scale in Rural Primary Education' in Curry, N. (ed.) (1981)
Department of the Environment (1982), *Structure Plans Advice Note 8/72* (HMSO, London)
Dorset County Council (1980), *Dorset (Excluding South-east) Structure Plan: Draft Written Statement*
Drudy, P.J. (ed.) (1976), *Regional and Rural Development* (Alpha Academic, Chalfont St Giles)
Dungate, M. (1980), 'Rural Self-help', *Voluntary Action*, 2, 17-21
Dunsire, A. (1978), *Implementation in a Bureaucracy* (Robertson, Oxford)
Fudge, C. (1981), 'Self-help and Social Policy', *The Planner*, 67, 60-1
Gilder, I.M. (1979), 'Rural Planning Policies: an Economic Appraisal', *Progress in Planning*, 11, 213-27
Gilder, I.M. (1981), 'Comments on the Economic Evaluation of Scale Economies in Rural Settlement Planning Strategies' in Curry, N. (ed.) (1981)
Gilder, I.M. and McLaughlin (1978), *Rural Communities in West Suffolk* (Chelmer Institute of Higher Education, Chelsmford)
Glasson, J. (1979), 'The Nature and Teaching of Implementation in Regional Development Planning' in Minay, C. (ed.) (1979)
Green, R.J. (1980), 'Planning the Rural Sub-regions: a Personal View', *Countryside Planning Yearbook*, 1, 67-85
Harman, R.G. (1978), 'Retailing in Rural Areas: a Case Study in Norfolk', *Geoforum*, 9, 107-26
Harrison, J.D. (1977), 'Geography and Planning: Convenient Relationship or Necessary Marriage?', *The Geographical Survey*, 6, 11-24
Healey, P. (1979), 'On Implementation: Some Thoughts on the Issues Raised by Planners' Current Interest in Implementation' in Minay, C. (ed.) (1979)
Hereford and Worcester County Council (1976), *Herefordshire Structure Plan: Approved Written Statement*
Hereford and Worcester Rural Community Development Project (1978), *Report of the Working Party* (Hereford and Worcester County Council)
Hereford and Worcester Rural Community Project (1980), *Schools Study 1977-79* (Hereford and Worcester County Council)
Hereford and Worcester Rural Community Development Project (1981), *Wyeside Community Project 1978-80* (Hereford and Worcester County Council)
Hertfordshire County Council (1976), *County Structure Plan: Written Statement*
Hodge, I.D. and Whitby, M. (1981), *Rural Employment: Trends Options, Choice* (Methuen, London)
Jenkins, W.I. (1978), *Policy Analysis: A Political and Organisational Perspective* (Robertson, London)
Kingham, M. (1981), 'Create Your Own Housing', *The Planner*, 67, 68-9
Kirby, D., Olsen, J.A., Sjφholt, P. and Stolen, J. (1981), *The Norwegian Aid to Shops in Sparsely Populated Areas* (Norwegian Fund for Market and Distribution Research, Oslo)
Knox, P. and Cottam, B. (1981), 'Rural Deprivation in Scotland: a Preliminary Assessment', *TESG*, 72, 162-75
Lefebvre, H. (1978), 'Reflections on the Politics of Space' in Peet, R. (ed.) (1978)
Levitt, R. (1980), *Implementing Public Policy* (Croom Helm, London)
McLaughlin, B.P. (1979), 'A New Role for Rural Community Councils', *Town and Country Planning*, 48, 124-5
McLaughlin, B.P' (1981), 'Rural Deprivation', *Town and County Planning*, 67, 31-3
Minay, C. (ed.) (1979), *Implementation – Views from an Ivory Tower*

(Department of Town Planning, Oxford Polytechnic)

Minay, C. (1979a), 'Four types of Planning Implementation' in Minay, C. (ed.) (1979)

Mitchell, B. (1979), *Geography and Resource Analysis* (Longman, London)

Morley, S. (1981), 'Positive Planning and Direct Development by Local Authorities', *Town Planning Review*, 52, 298-306

Moseley, M.J. (1979), *Accessibility: The Rural Challenge* (Methuen, London)

Moseley, M.J. (1980), 'Rural Development and its Relevance to the Inner City Debate, SSRC, Inner Cities Working Paper No. 9

Moseley, M.J. (1980a), 'Is Rural Deprivation Really Rural?' *The Planner*, 66, 97

Moseley, M.J., Harman, R.G., Coles, O.B. and Spencer, M.B. (1977), *Rural Transport and Accessibility* (Centre of East Anglian Studies, University of East Anglia)

Norfolk County Council (1976), *The North Walsham Area* (Norfolk County Council, Norwich)

Northamptonshire County Council (1977), *County Structure Plan: Report of Survey*

Nutley, S.D. (1981), *The Evaluation of Accessibility Levels in Rural Areas* (Welsh Office, Cardiff)

Olsson, G. (1978), 'Servitude and Inequality in Spatial Planning: Ideology and Methodology in Conflict' in Peet, R. (ed.) (1978)

O'Riordan, T. (1971), *Perspectives on Resource Management* (Pion, London)

Parsons, D. (1979), 'A Geographical Examination of the Twentieth Century Theory and Practice of Selected Village Development in England', unpublished PhD thesis, University of Nottingham

Pearce, J. and Hopwood, S. (1981), 'Create Your Own Jobs – 2. Community Based Initiatives', *The Planner*, 67, 64-7

Peet, R. (ed.) (1978), *Radical Geography: Alternative Viewpoints on Contemporary Social Issues* (Methuen, London)

Pressman, J. and Wildavsky, A. (1973), *Implementation* (University of California Press)

Rawson, J. (1981), 'The Impact of Rural Settlement Policies of Concentration and Dispersal in Rural Britain', Paper presented to the PTRC Summer Annual Meeting, University of Warwick

Royal Town Planning Institute (1979), *Making Planning More Effective* (RTPI, London)

Runciman, W.G. (1972), *Relative Deprivation and Social Justice* (Penguin, Harmondsworth)

Russell, A.J. (1975), *The Village in Myth and Reality* (Chester House, London)

Shankland, G. (1981), 'Create Your Own Jobs – 1. The Great Uncounted: the Role of the Informal Economy', *The Planner*, 67, 62-3

Shaw, J.M. (1976), 'Can We Afford Villages?', *Built Environment*, 2, 135-7

Shaw, J.M. (ed.) (1979), *Rural Deprivation and Planning* (Geobooks, Norwich)

Simmons, I.G. (1974), *The Ecology of Natural Resources* (Edward Arnold, London)

Smigielski, K. (1978), *Self-Supporting Co-operative Village* (Building and Social Housing Foundation, Coalville)

Voluntary Action Cumbria (1974), *Rural Communities Project Report* (Voluntary Action, Cumbria)

Walker, A. (ed.) (1978), *Rural Poverty: Poverty, Deprivation and Planning in Rural Areas* (Child Poverty Action Group, London)

Warford, J.J. (1969), *The South Atcham Scheme: An Economic Appraisal* (HMSO, London)

Whitby, M.C. and Willis, K.G. (1978), *Rural Resource Development: An*

Economic Approach (Methuen, London)

Willis, K.G. (1980), *The Economics of Town and Country Planning* (Granada, St Albans)

Wilmers, P. (1981), 'Planning, Self-help and Mutual Aid', *The Planner*, 67, 59

Winter, H. (1980), *Homes for Locals?* (Community Council of Devon, Exeter)

Wolfenden Committee (1978), *The Future of Voluntary Organisations* (Croom Helm, London)

Woollett, S. (1981), *Alternative Rural Services* (National Council for Voluntary Organisations, London)

Working Party on Rural Settlement Policies (1979), *A Future for the Village* (HMSO, Bristol)

10 RURAL PLANNING

D.L.J. Robins

The rationale of rural planning systems lies in the inability of society to secure a coherent and stable pattern of land use in rural areas in such a way as to produce the most life-enhancing pattern of activity for the community as a whole. In broad terms, therefore, planning is a form of public intervention to secure this end. It is a statutory function, having the consent of society for its legislative force. Although, superficially, rural planning systems overtly represent the systematic allocation of land to particular uses, in another way they represent the resolution of competing and conflicting value systems in society. Rural areas with a close-grained pattern of settlement, as in much of Britain, where there is a high level of differentiation within short distances, are characterised by a multiplicity of issues, interests and authorities. Conflicting demands arise from a wide variety of potential private and public activities reflecting social, economic and aesthetic aspirations. The kinds of conflict which need to be resolved through the planning process are not, however, simply between public and private interests, with the state representing as referee the community at large. Conflicts, also arise between private interests — such as between private forestry and agriculture — or even between different agencies of government, as in the current debate on the conflict between public amenity objectives and state support for the agricultural industry. Over a period of time, rural landscapes may be seen as a palimpsest demonstrating those successive outcomes of conflicts over the use of land whose advocates have been successful against other claimants. Without a planning system, they merely demonstrate whose preferences are sustained by an unrestrained trial of physical or economic strength. Within a planning system, on the other hand, they give effect to the preferences of society as a whole. The planning system has the function of a public broker or referee.

Plans clearly have to conform to some kind of formal specification. Broadly, any planning scheme must identify the authority by which it is produced; the agency responsible for its production; its functional areas of concern, both geographically and in terms of the issues to which it relates; the future state of the activities to which it relates,

226

which must be measurable in acceptable terms; and it must identify the resources available for utilising its ends. It is not a once-and-for-all but a cyclic activity, subject to continuous revision on a rolling programme.

Any planning system acting as a basis for public decision-making must be clear about the issues with which it is dealing; must have access to adequate information about the issues; and must be able to measure the effects of public and private decisions about those issues. The issues will concern the realisation of goals of an economic, social or aesthetic nature about the best use of land. The system therefore has the overwhelming characteristic of reducing uncertainty about the future, in the light of goals which are normative rather than positive.

Physical planning in general has been the subject of a great deal of conceptual and theoretical development since the 1960s. The end of this activity has been to raise planning from the status of a pragmatic action-oriented operation with vague, ill-defined goals, to one supported by coherent theory, itself subject to continuous development and criticism. However, the geographical base of this work has normally been concerned with urban issues and rural planning still lacks a cohesive philosophical base (Davidson and Wibberley, 1977). The traditional idea of a rural-urban settlement continuum militates against the notion that rural planning problems are in any way distinctive and may even inhibit the perception of rural development and management systems. The reasons for this lie in the fragmentary and often marginal nature of decision-making on rural development, in which the inter-relatedness of the work of a multiplicity of decision-making bodies is only slowly being perceived. It is not therefore surprising that a developed country such as Britain demonstrates the need for a means of resolving competing land-use demands — yet the search for common goals, objectives and standards is elusive. An evaluative process common to physical, biological, visual and managerial problems in rural areas is required but, as Davidson and Wibberley point out, a monetary common denominator may not be universally appropriate as the basis of choice and may unjustifiably supplant other measures of resource attractiveness. Cherry (1976) argues that land-use analysis alone is insufficient and that planning should establish common measures but that a social base would be more appropriate for selecting alternative rural planning policies.

Rural Issues

The broad issues of planning for rural areas are effectively spelled out by Blacksell and Gilg (1981), who point to four geographical type areas, each of which has distinctive planning problems. Firstly, there is the urban fringe, that rural buffer against which urban expansion must collide. This area has to satisfy an urbanising population's demand for space in the context of the use of increasing wealth to acquire space and greater personal mobility. For this reason, urban containment has been a durable goal in British planning (Hall *et al.*, 1973). Nevertheless, while cities may have a relatively firm bricks-and-mortar edge under such policies, neighbouring agricultural land is highly subject to the influence of urban activities.

The broad result of urban containment has been more coherent urban structure although it has not prevented the conversion of agricultural land to urban uses. While ribbon development has now been effectively contained for almost fifty years, peripheral housing estates, new towns and expanded towns have made continual demands upon rural land. Nevertheless, abatement of the scale of this demand, tested by research in several different areas, is reported by Anderson and Best (1981). They cite work by various researchers demonstrating that in London and the South East of England urban physical expansion has slowed markedly in the last two decades and that a rising population has been accommodated within existing urban areas by increasing residential densities and in other areas by developing land formerly in non-agricultural use. Other work, on Tyneside, showed a larger proportion of agricultural land-take and a similar study in East Sussex showed that in 1975–6 farm land comprised 53 per cent of the land for which planning permission was sought but only 31 per cent of the land for which permission was granted. As Anderson and Best imply, the public view that substantial and rapid losses of agricultural land are occurring is curiously persistent when research results are indicating otherwise.

While in many ways urban containment has been a consistent and potent mainstream in British post-war planning, erosion of the rural areas has persisted since the redevelopment of nineteenth-century inner residential areas of cities necessitated a lowering of densities and the consequent creation of a housing demand which had to be met elsewhere. Residential development has not been the only demand, however. The structure of modern industry with the large single-storey factory set in spacious grounds also looks to the urban periphery for sites. The case for the structuring of urban growth was well founded, but however urban expansion is controlled there is always a zone of peripheral open land very much affected by the town's proximity. Trespass and theft of crops, carelessness in leaving gates open and damage to fences are obvious signs of a nearby urban population. The fringe area in these circumstances takes on a quite different character from that of secure agricultural land; short-term cash cropping, pig keeping, or the keeping of horses for leisure activities are typical manifestations.

The second type of rural zone having coherent planning problems is that of the agricultural lowlands. In post-1945 Britain the ability of urban income earners to meet the cost of longer journeys to work, especially by car, has resulted in what would otherwise have been a high proportion of peripheral urban development taking place in rural settlements instead. This movement has implanted urban life-styles in villages; it has on the one hand partnered the expansion of non-personal physical services such as electrification and sewage disposal but on the other hand has been accompanied by a widespread decline in such personal services as public transport or retailing since the ex-urbanite uses his car to gain access to city services. The losers tend to be the indigenous population employed locally in primary industries. As well as losing services they find themselves also in a more highly competitive housing market. The changing technology of agriculture itself has raised concern, particularly in the areas of extensive arable farming in lowland Britain, where tree and hedge removal and changes in the field pattern in the interests of production economies have effected substantial landscape change outside the scope of normal planning control.

The third and fourth zones of Blacksell and Gilg, the hills and remote uplands, have rather different problems. Socially areas of declining indigenous population, low employment opportunity, and low incomes, they are nevertheless important as an arena for recreation, on a regional or national scale. However, they are also service areas for the nation as a whole in a number of other respects. They provide water supplies, largely consumed in urban areas, from river intakes and impounding and balancing reservoirs; they are the source of various minerals; and they provide suitably wild and comparatively unpopulated areas for defence training. Decisions on land use concerning these matters have generally been taken at well above the local level. Indeed, since national populations in developed countries are in general overwhelmingly urban the strength of these land-use demands is, arguably, primarily assignable to urban needs.

Two other problems in these zones require mention. In the more attractive areas of the uplands, the open countryside, or at the coast, the affluent urban-dweller seeking a second home can upset the indigenous housing market. Secondly, like the lowlands these zones are also affected by changes in the primary industries in ways lying outside planning control. The ploughing up of moorland for arable and competition between agriculture and forestry — too often soft-wood monoculture — have given rise to concern about landscape changes.

The broad conclusion of Blacksell and Gilg is that two planning approaches are at work concurrently. The statutory land-use planning system working in a rather negative way with its substantial rights of veto over building is accompanied by a rather diverse medley of resource planning functions covering such interests as agriculture, forestry, recreation and tourism. These functions are not necessarily convergent.

The Institutional Framework

The statutory planning system of England and Wales dates in its present form from 1948. Scotland is subject to independent and broadly similar arrangements. The original main authority for the system was the Town and Country Planning Act, 1947, born in a welter of legislation in the post-war reconstruction period. Planning before 1948 is now the subject of historical study; its legislative form is well traced by Heap (1978) and its technical form by Cullingworth (1976). The system was created in response to three notable wartime reports, known popularly by the names of their respective chairmen: the Barlow Report (1940), dealing with the distribution of the industrial population and the disadvantages of its concentration; the Uthwatt Report (1942), which analysed problems of the payment of compensation and recovery of betterment in respect of the public control of land use within the wider problem of the terms on which land could be acquired for public uses on an equitable basis; and the Scott Report (1942) on Land Utilization in Rural Areas with terms of reference 'to consider the conditions which should govern building and other constructional development in rural areas, consistently with the maintenance of agriculture, and in particular the factors affecting the location of industry, having regard to economic operation, part-time and seasonal employment, the well being of rural communities and the preservation of rural amenities'.

It is worth noting the Scott Report's recommendations, particularly as their diversity was such that they were not all appropriate to be incorporated in the 1947 Act planning system though they very fully reflected the variety of rural issues and public responsibilities for them. Scott called for improvements in the design standards of public rural housing and a reduction in the influence of the 'tied' cottage. Better provision of village halls and playing fields would stimulate rural social life. The nascent mobility of the urban population and its need for access to the countryside without detriment to agricultural interests were recognised. National parks, nature reserves, and holiday camps were advocated. The views of the Davidson Committee's review of the Ordnance Survey's tasks were taken into account in supporting the production of a new map at the scale of 1:25,000 which would be of great use in rural planning. Local planning administration needed compulsory powers within the framework of a strong central co-ordinating authority.

Two further reports were especially significant for the development of rural planning. The Dower Report (1945) established the desirability of a national park system. The Hobhouse Report (1947) went further in considering areas suitable for designation.

While the 1947 Planning Act provided the overall 'umbrella' for planning control certain other legislation was also significant, namely:

(i) The New Towns Act, 1946 followed the Barlow recommendations and Abercrombie's 1944 Greater London Plan and proposed a green belt ringing London. In Abercrombie's plan it constituted the third of four concentric rings comprising Greater London. Its function was to contain the metropolis by the prevention of development. After 1948 it became safeguarded — with reviews of its area from time to time — through planning control rather than by public acquisition as foreseen under the previous Green Belt (London and Home Counties) Act, 1938. The New Towns, built by Development Corporations funded directly by the Treasury under the 1946 Act, were built in Abercrombie's fourth ring, furthest from London, and sites were chosen or restricted with regard to agricultural interests. Other New Towns were built for rather different reasons in other regions (Cullingworth, 1976).

(ii) The Town Development Act, 1952 enabled existing local government authorities to provide for the relief of population congestion in the conurbations by the development of provincial towns in partnership with large city authorities. By this means substantial development occurred in many otherwise fairly static small towns such as Thetford (Norfolk), Haverhill (Suffolk) and Huntingdon.

(iii) The National Parks and Access to the Countryside Act, 1949 was immensely important for rural planning, constituting the legislative sequel to the Dower and Hobhouse Reports and some of the Scott Report's issues. It designated as national parks all the areas considered by Hobhouse, with the exception of the Norfolk Broads; these were defined in Dower as extensive areas of beautiful wild country in which in the national interest the characteristic landscape beauty was to be protected; access and open-air recreation facilities were to be provided; and the indigenous farming economy was to be maintained.

National park authorities were not, however, set up independently as Hobhouse had advocated and the parks were administered by the local planning authorities under a special committee. At central government level the National Parks Commission, a relatively small body with advisory powers only, was created by the 1949 Act.

At a lower level of national significance than the national parks Hobhouse identified some fifty-two high-quality landscapes. The Act conferred powers on the National Parks Commission to designate these as Areas of Outstanding Natural Beauty, administered by local planning authorities and qualifying for central government financial assistance for work protecting or enhancing their natural beauty; thirty-three were designated by 1980.

A further important provision of the 1949 Act was the creation of the Nature Conservancy to give advice on the conservation of sites of biological or geological interest. It was also to foster research and establish and manage nature reserves, as could local planning authorities. Under Section 21 of the Act, localities could be defined as sites of special scientific interest.

Yet another element of the 1949 Act's provisions was the requirement for county councils to make footpath maps — legal and incontrovertible statements of rights of way, established after a lengthy and formal but democratic procedure. These have become incorporated in Ordnance Survey maps as their status has become established. The National Parks Commission and its successor the Countryside Commission worked simultaneously on the establishment of long-distance routes.

The 1947 Development Plan System

The 1947 Act was a milestone in planning. Its broad provisions were:

(i) To define county boroughs and counties as the local planning authorities, so that about 1,400 former authorities in England and Wales were reduced to a tenth of that number; the Minister at central government level was charged with co-ordinating their activity;

(ii) the requirement for development plans to be formulated for land use for the next 20 years at a scale of 1″ to the mile for counties, 6″ to the mile for urban areas and 25″ to the mile for a Comprehensive Development Area, generally a small, usually urban, tract requiring re-development probably under compulsory purchase powers;

(iii) the requirement for all 'development' to have consent, development constituting all but insignificant forms of construction, engineering and mining and any material change in the use of land or existing buildings;

(iv) an attempt to solve the compensation betterment problem investigated by Uthwatt;

(v) highly specific provision for the conservation of amenities.

The 1947 style of plan served for more than two decades. Nevertheless, its formulation was subject to a cumbersome formal process and its production and approval were excessively delayed in many cases. Its definitive element, the *Written Statement* contained in a typical county, a curious and uncorrelated schedule of proposals. On the map, the main road pattern, areas of special landscape policy, mineral workings and Sites of Special Scientific Interest were typically shown; the map also served as an index for the larger-scale town maps. It also usually showed key rural settlements with a diagrammatic representation of public facilities proposed to be provided.

Consequences of 1947 Plans

In the long run, however, while the 1947 style plan was a successful instrument for controlling undesirable development it was not successful as a vehicle for expressing the future of the rural economy in land-use terms; it was too fragmentary and omitted too many significant influences for that.

A clear result of the form of the 1947 plan was that most of the land shown on a county map bore no notation at all; it was colloquially referred to as 'white land' — an area where existing uses would remain undisturbed unless carried on in contravention of planning control; great weight therefore attached to day-to-day development control decisions. Also, because many rural towns were too small to warrant the production of town maps there arose a practice of formulating informal plans with varying degrees of non-statutory significance.

Particularly crucial was the status of agricultural and forestry operations. These were generally exempt from planning control save in the case of agricultural buildings close to roads, of large floor space, or exceeding a defined height. These exemptions 'caught' such modern rural features as the broiler house or tower silo within the net of planning control. The general result of this was that the two primary rural industries enjoyed a degree of freedom from planning control quite unlike other forms of economic activity. This exemption has become a matter of grave concern for reasons we examine later.

The least stable element of the 1947 Act was its financial arrangements. The compensation betterment issue has proved an Achilles heel of planning, subject more than any other part of the planning system to the vagaries of political change.

The Countryside Act 1968 was a significant forward step for rural planning. The Countryside Commission, replacing the National Parks Commission, became a highly important body in rural planning in general. While the Commission does not own or manage land — save for temporary experimental purposes — its advice and encouragement have been fundamental to the development of rural planning, through studies of particular issues, such as its work on Heritage Coasts (Countryside Commission, 1970a, b), research into particular problems, and the conduct of specific experimental work in rural planning (Hookway, 1977).

The Planning Advisory Group and the 1968 Town and Country Planning Act

The growth in the production of informal plans and the growing inadequacy and imprecision of the formal plans, together with the widespread feeling that the public at large had scant part in plan-making led in the 1960s to the appointment of two commissions. The Planning Advisory Group was appointed in 1964 to 'assist in a general review of the planning system'. It reported in 1965 and its recommendations (Planning Advisory Group, 1965) were largely adopted in the Town and Country Planning Act, 1968. Another group, commonly referred to as the Skeffington Committee, looked into the question of public participation, reporting in 1969 (Skeffington Report, 1969). Again, the 1968 Act reflects their deliberations.

The 1968 Act established the responsibility of central government by reserving to the Minister the final authority in policy and strategic issues to be published as a Structure Plan prepared by the local planning authority. The local authority deals with detailed local issues in local plans formulated within the structure plan framework. These may comprise district plans, general land use plans, or subject plans, for a highly systematic purpose such as recreation or reclamation. Action area plans deal specifically with major change in small areas. Unlike their predecessors, the Comprehensive Development Area maps, they do not necessarily imply compulsory purchase and public development.

Structure Plans are intended to be based on realistic forecasts of resources in the public sector and hence should reliably indicate the major policies and objectives for planning in their areas. They abandon the traditional Ordnance Survey base precisely because they are about policies and not about the future of individual properties, which can only be identified in local plans which retain their location-specific base maps. In other words, the illustrative material of structure plans may be highly varied and in addition to the text include diagrams, tables and 'broad brush' cartography. The plans are to be reviewed as necessary and not at set intervals as in the case of 1947 Act plans; this brings much needed flexibility.

Local plans, on the other hand, are intended to afford positive guidance to developers and the development control function. Action area plans in a rural area, prescribing comprehensive development on a ten-year horizon, provide for highly specific activity such as a mineral workings restoration plan or the development of a country park.

The re-organisation of local government in England and Wales in 1974 created a two-tier system of planning administration. In rural areas, counties are responsible for Structure Plans and district authorities for local plans formulated within the Structure Plan framework. The downturn in economic vitality in the latter half of the 1970s tended to withdraw attention from Structure Plan making with a concomitant reduction in the saliency of county councils in planning and to bring greater emphasis to the tactical work of district planners on local plans.

Public Participation

The post-1968 planning system envisages an ongoing commitment to public participation in plan-making and especially in the formulation of Structure Plans. The authority must give publicity to the issues proposed to be included in the Structure Plan and to the contents of its own surveys. The public must be afforded the opportunity to make representations and the authority must take these into account in drawing up its plan. The draft plan must be available for public inspection and objections may be lodged with the Minister. At the same time the authority must notify the Minister of steps taken to publicise the plan-making process and to deal with representations about the proposals. If he is not satisfied, the Minister may refer the plan back to the authority for them to afford it better publicity. Finally, the Minister orders an examination in public, by a small panel sitting with an independent Chairman, of issues which central government considers important.

In the outcome, this innovation has not been very useful in improving the quality of plan-making. The problem is that once participation becomes a consideration of principles the traditional type of objection, made on the ground of interference with individual property, falls away. Experience has demonstrated the difficulty of public response to high-level statements of general policy.

A further difficulty in rural areas in particular is that of identifying the public in whose interest the plan is made. Ideally, one would hope to consult informed and representative interests; yet since participants are likely to be impassioned they will not be disinterested as well. Obvious conflicts between the interests of rural residents, individual non-residents who enjoy rural areas for recreation, and the large-scale land-use demands of water supply, afforestation, mineral needs and defence are not easily reconcilable. Whereas some representations will be based on ideals, most will still be based on wants.

Recent Developments

Britain has no national physical development plan as such. The Secretary of State for the Environment together with the Secretaries of State for Wales and Scotland have the overall responsibility for ensuring the consonance of Structure Plans in a national setting but these are not formally conflated into a national plan. When the 1968 Act system was devised the Structure Plans were intended to be co-ordinated by the Regional Economic Planning Boards. This link was never very strong, however, since the Boards only had advisory rather than executive functions. They were totally abolished in changes following the 1979 General Election.

The notion of comprehensiveness in Structure Plans is modified to some extent in the light of recent ministerial treatment of their social content (Jowell and Noble, 1982). Whereas the Planning Act requires that social and community services be considered along with policies for the development and use of land and this is reinforced by the Department of the Environment Circular 4/79 calling for 'broad and open consideration of physical problems taking account of physical, economic, and social aspects' recent ministerial response to submitted plans has indicated limitations. While discussion of broad social strategies has been upheld by central government, actual social policies have been removed on submission of plans to the minister. Jowell and Noble indicate three kinds of policy which have been excluded: social policies requiring the alteration of central government's policies or administrative arrangements; policies based on users rather than uses, as in the case of the attempt to control housing policy in the Peak District; and policies to benefit particular client groups through positive discrimination in the allocation of resources. This is clearly a setback for hopes of structure planning based on social considerations.

This tendency is reflected in Martin and Vorhees Associates' (1981) review of thirty-five years of rural settlement policies in Britain carried out for the Department of the Environment. They draw an important distinction between those rural areas affected by the immigration of people with urban life-styles and the remote rural areas affected by general depopulation. Settlement policies are held to be important in the former case but in the latter situation they are of far less import than economic and social development in an aspatial sense. In the remote areas Martin and Vorhees believe the providers of social services and transport facilities are more significant than planners. Examining a variety of key settlement approaches in such areas, they find the planning policies of slight impact and incapable on their own of arresting decline. They advocate detailed study of rural change, particularly of cost implications and access to services, and sympathy on the part of planning authorities to proposals which would improve rural facilities.

While Structure Plans have not entirely fulfilled the expectations held for them in 1970, local plans have also encountered difficulty. Under present arrangements the local planning authority is expected to consult central government about the need for local plans. Central government's view, however, is nebulous, for there is no general rule about the need for local plans and they are not considered necessary where little or no pressure for development is expected. Such advice unduly emphasises 'development' − in the sense of bricks and mortar − when the diversity of issues in rural planning is only partially concerned with building, and is likely to result in the kinds of non-statutory plan which PAG and the 1968 Act sought to eliminate.

Experimental Work and National Park Developments

That problems of rural planning do not align with the hierarchy of national parks, areas of outstanding natural beauty, and the rest of rural Britain is shown by recent examples of experimental work in a variety of very different types of locality.

A Countryside Plan devised by Essex County Council (1981), for the use of its constituent authorities as supplementary planning guidance, employs a categorisation system of areas of predominant interest in terms of areas for agriculture, zones for conservation, special landscape areas and landscape improvement areas. The significant aspect of the policy's approach is that it stands on the voluntary principle rather than the minutiae of Planning Act legislation. Such an approach is thought likely to induce local initiative and even financial support. A very different problem is described by the Countryside Commission (1981) in two areas of the urban fringe of Greater London where woodland rehabilitation and tree planting were undertaken and access to the countryside was improved by attention to footpaths and bridleways. In these zones the danger of damaging trespass is prevalent and there is much virtually derelict land. Here again, improvement has been achieved slowly through small-scale project work in which management and implementation rather than planning are the keynotes. Landscape of a quite different quality forms the subject of Wager's (1981) work for the Peak Park Planning Board on historic landscape in a national park. The problem here is that erosion, neglect, arable farming and forestry are obliterating what we might regard as agricultural archaeology — features such as field systems and mounds which are not necessarily sufficiently significant to be protected by the Ancient Monuments Act. While powers to acquire land exist the national park movement in Britain has never been predicated on extensive land purchases but rather on an ethos of control to which Wager adds the spirit of persuasion through landowners' voluntary agreements.

In the National Parks, management plans devised following the publication of the DOE Circular 65/74 have spelled out the basis of negotiated agreements between rural interests. The plans have their roots in effective public participation and deal in a realistic way with physical issues of immediate concern. Indeed, their remit contrasts strongly with the structure planning system in that they deal with small-scale short-term issues. Hookway (1977, 1978) claims that they are more serviceable than Structure Plans in resolving issues leading to conflict between agricultural, forestry, water, recreation, and tourist interests. The plans should essentially be seen as instruments dealing with issues unforeseen in 1949. The parks were designated to maintain and enhance the natural beauty of their landscapes; and the indigenous farming pattern was to be safeguarded. The national parks were, by and large, sited in relatively remote upland or coastal areas. Recreation was likely to be of types attractive to the few rather than the many, active rather than passive, and in general consonant with the relative wildness of the landscape. The Peak District was perhaps always exceptional compared with other national parks in its accessibility for large con-urban populations around its fringe. Nevertheless, the scale of access by private transport to areas conceived of as fairly remote in 1949 was considerable three decades later. For the Lake District in particular the construction of the M6 motorway as a new major national north-south route brought the Park within day-trip accessibility of a large proportion of the population of Britain. This population was not merely participating in the relatively strenuous and, within broad limits of park capacity, active outdoor forms of recreation appropriate in 1949, but was also tending to bring into the park a car-borne visitor interested in sufficient parking spaces, preferably at beauty spots, and designed in such a way that he might not have to alight from his car. The Goyt Valley experiment in the Peak District offers a good example of the use of official minibuses to frustrate car penetration. Neverthe-less, the contribution of the National Parks as a major rural 'lung' for the urban population, as originally seen by the Dower Report (1945), has not been achieved. This is particularly the case in the South East where the public are therefore forced to seek rural recreational areas of a quality different from the parks.

While in general the agricultural landscapes of the parks are less under threat than those of lowland areas since they are unsuitable for large-scale arable operations they have been subject in places to the search for more productive methods of husbandry. A particularly conspicuous case has been the ploughing up of the marginal moorlands of Exmoor; the paradoxical feature here has been the change from uneconomical to more profitable farming accompanied by a very marked change in the visual quality of the landscape. There is a clear need for a categorisation of landscape quality with a view to designating those landscapes most urgently requiring conservation. The National Parks Review Committee (Sandford Committee) in 1974 conceived the idea of National Heritage Areas as a designation for the tracts of highest landscape value. The Countryside Review Committee (1979) was of the view that a two-tier classification of rural landscape should be made.

The higher quality, embracing perhaps 1 to 2 per cent of rural land, might be subject to acquisition in the public interest. The proposal itself was unusual, given that the British rural planning system favours control measures rather than public ownership. In the second tier of the classification would be land suitable only for a lesser conservation priority and typifying much of the land area of the present parks. The problem of such a system would be its greater complexity than the current undifferentiated subject. The higher quality areas might inevitably, because of their status, defeat their purpose by being self-evidently more attractive. Significantly, the Sandford recommendations were not adopted by central government.

The general trend in planning towards emphasis of the distributional effects of plans in social and economic terms raises a large and significant new issue for the parks. Socially, they are areas of depopulation and ageing populations, and of emigration of the younger population on account of the lack of employment opportunities. Marginal types of hill farming in unviable units and a general lack of enterprise serve to restrict people's life chances. To some extent the national park is a positive attraction to development through the tourist industry in the development of hotels and more informal accommodation such as bed-and-breakfast facilities, or camping sites. These features, however desirable in underpinning the local economy, are precisely those which cause difficulties in terms of the original objectives of park designation. Recent work at Edinburgh University which has drawn attention to this in detail (TRRU, 1981) has pointed out the need for specific national guidelines to replace present *ad hoc* approaches to the conservation of scenic beauty and for national park planning to encourage the retention of a viable local economy with a stable resident population. The Unit points out that economic activities, with the exception of support given by the Tourism Act of 1969 to appropriate activity, receive little direct public aid. The bringing under planning control of agricultural and forestry operations is advocated on the grounds that these activities are capable of bringing material change to an area. Such an administrative change would, however, have to take place in conjunction with a fundamental remodelling of the grant system of the Ministry of Agriculture and the provision of its advisory services in a way which would support congruent policies of conservation, recreation provision, and the support of the agricultural labour force. Another recent review of national park policies is not so confident about the potential efficacy of planning control (MacEwen and MacEwen, 1981). Whereas the MacEwens are in favour of the extension to control all building,

Printed and bound by CPI Group (UK) Ltd, Croydon, CR0 4YY
01/11/2024
01782626-0002

engineering or land-drainage works associated with agriculture or forestry, they have strong reservations about extending control to 'living systems'. Their argument is that planners are not equipped to extend their range of expertise beyond the built environment and that management schemes, backed by the power of compulsion, should be devised by officers with appropriate skills and knowledge in consultation with the farming and forestry interests.

The issue on which both reports converge is the need for a change of outlook in the agricultural industry and the Ministry of Agriculture's concept of its duty to support a primary function of production. How far this can be consonant with a conservation objective is an important *raison-d'être* of the Wildlife and Countryside Act of 1981, discussed below.

Settlement Policy

One common feature of earlier styles of plan was the 'key settlement' approach, based essentially on a rationale of economies of scale in the provision of public services. This commonly selected key settlements for development and is still the basis of much rural structure planning. More recently it has come under criticism (Clarke, Ayton and Gilder, 1980) as 'a cosmetic justification for a policy created merely out of economic expediency and administrative pragmatism'. Ayton in the same paper defends the principle but Gilder asserts that the concept is only valid for a limited range of services. A few structure plans, notably those of North Yorkshire, Gloucestershire and Cumbria, are based on service thresholds appropriate to a 'necklace' of five or six villages rather than designating a key village for development and public investment (Derounian, 1980). This may herald the end of an established shibboleth in British rural planning, though it is perhaps too early to see moves at the local level, such as the self-help minibus or a development control policy favouring village stores and post offices, becoming general practice. In any case, many small scale activities of this kind depend on initiatives other than those of the planning authority (Shaw, 1980).

Rural Deprivation

An issue which must increasingly enter rural planning considerations is that of rural deprivation. The recognition that rural areas may contain substantial social deprivation reflects Cherry's (1976) call for a social base to rural planning which is quite distinct from the recreational and conservationist emphases of earlier decades. While no single measure of poverty is exclusively characteristic of rural areas a combination of factors renders rural life less attractive to the housewife, the young, the aged, and the sick. McLaughlin (1981) admirably summarises the problem with reference to a large volume of recent work on the subject. There is an imbalance for poorer and less mobile rural inhabitants of what are now fashionably entitled 'life chances'. In real terms, rural employment is characterised by low wages, low levels of skill, poor opportunity for employment and long-term high levels of unemployment. The job market offers little, therefore, to married women, the unskilled, or school leavers. The housing market, because of the rural spread of commuterdom and second homes, works in favour of the ex-urban adventitious inhabitant rather than the indigenous population dependent on primary industry. There is a crisis of accessibility; dispersed and often declining populations can only be serviced at increasing cost or with gradually poorer service delivery, perhaps to the point of complete withdrawal. The prevailing political influence in rural areas favours private enterprise rather than public intervention.

However, though there are clear socio-economic similarities between inner city deprivation — the focus of much current attention in Britain, particularly after the city riots of the 1981 summer — and rural area deprivation, the rural situation is different politically, geographically and environmentally. Does its symptomatic commonality with urban deprivation demonstrate its aspatial character as claimed by Moseley (1980)? As he points out, no sophisticated managerialist or political models of rural decision-making throwing light on the producers rather than the consumers of rural deprivation have yet been derived, as they have in the case of urban poverty. Such a conceptual development is needed and its relation to spatial issues has to be understood if it is to become a main thread in rural plans in the 1980s.

Rural Employment

The long-sustained concern with conservation and visual amenity in rural planning has now highlighted in a negative way considerations about employment in the countryside. Several years of national economic depression have thrown into focus the potential balance between employment opportunity and the maintenance of amenity. Indeed the Department of the Environment in Circular 22/80 *Development Control: Policy and Practice* emphasises the need to balance conservation considerations against encouraging the provision of employment. This has become a concern of central government in view of the traditional assertions of the business world that planning permission is often witheld from enterprises in rural areas; the planning system is subject to undue administrative delay; and the grant of planning permission is often accompanied by onerous conditions (Taylor and Bozeat, 1982). Appropriate job-creating activities in rural areas to counteract the decline in the labour demands of primary industry are suggested in a Labour Party (1981) policy document. Suitable industries would be connected with food processing, furniture-making and general craft work; with the scientific instrumentation, particularly the microprocessor, representing small-bulk, light products with high added value in processing; and agricultural engineering. These developments are reckoned necessary as an antidote to the traditional idyllic picture of rural areas cherished by landowning and farming interests. Reflection demonstrates, however, that the only truly original suggestion is the development of microprocessor-based industry; this must be the archetype of footloose industry, for which both urban and rural locations alike must compete. On the other hand, rural areas may possibly have the advantage in this competition where they offer attractive physical surroundings.

Rural Housing

While the provision of housing for the indigenous rural population was a concern of the Scott Committee the subsequent forty years have seen rural housing problems develop in ways which were not significant until comparatively recently. Broadly, the availability of rural housing for local people has been affected by the demand for existing houses of those moving from urban to rural areas; by the growth of the second-home movement, essentially weighing on rural areas; and by recent legislation enabling occupiers of local authority housing to purchase their houses at highly advantageous prices, thereby taking their dwellings out of the public housing stock.

A particularly contentious issue is central government's powers under Section 19 of the Housing Act, 1980. Whereas the Act gave effect to the tenant's right to buy local authority housing central government retained the power to designate areas in which subsequent resale could be controlled by the local authority, to the end that any particular dwelling might be repurchased for return to the public stock. The provision is ingenious in its attempt to reconcile choice and need and in the event 130 English authorities applied for designation (Clark, 1981). National parks and areas of outstanding beauty were automatically scheduled but otherwise only eighteen authorities were accorded powers over resale. While these included Cornwall and much of rural Northumberland, substantial areas of the London green belt and such regions as the Cotswolds and East Anglia failed to obtain powers yet these are precisely the types of area where the incomer and second-homer are active in the market to the disadvantage of local inhabitants with fewer financial resources. Unsurprisingly, the Labour Party (1981) advocates the use of Section 52 agreements under the Planning code to compel the occupancy of new housing by local residents. Restrictive planning policies on new development and the sale of rural council housing have exacerbated the rural housing problem (National Federation of Housing Associations, 1981).

Planning Control of Agriculture?

The British statutory planning system essentially concerns development and material changes of use. In essence these refer to the built environment and in many respects are based on urban issues. From the start of the system in 1948 agricultural practices largely enjoyed immunity from planning control. Buildings within a defined distance of a public road were subject to planning control and with the technical development of agricultural practice eventually all larger buildings – notably including broiler chicken houses and tower silos – came under control. However, changes in agriculture have now achieved considerable momentum in terms of the loss of hedgerows, trees, and woodlands, particularly in areas suitable for large-scale arable farming. The changes have been documented by Fairbrother (1970) and Westmacott and Worthington (1974). More recently Shoard (1980) has drawn attention to the problems. At the same time, at the moorland margin semi-natural vegetation has been ploughed for arable, most conspicuously in the Exmoor National Park (Parry, 1981). Many small-scale landscape features such as ditches and ponds once had functional significance in farming. Now they are unnecessary, neglected and do not contribute to the farm economy. Indeed, their maintenance would reduce profitability.

Clearly, the extension of the urban planning principles of development control or redevelopment to agricultural landscapes in the interest of visual amenity is an attractive idea. On the other hand, it is difficult to envisage its working for it is doubtful whether the requisite expertise could be afforded by planning authorities; the farming lobby is politically powerful; the monitoring and more especially the enforcement of conservation measures would present novel problems; and the aura of a greater degree of control would be likely to antagonise landowning and agricultural interests. Over the whole issue hangs the problem of compensation. If a farmer is prevented by planning control from making the most economic use of his land on what basis is he to be compensated for his loss? In terms of loss of annual profits or of depreciation in land value?

A more suitable but informal solution to the problems may lie in voluntary agreements in return for some form of compensation. Informality would not be hedged by statutory procedures; it would encourage local initiative and experimentation; and it could become the basis of goodwill between the competing interests. However, the scale of operation required would be so large in some areas that it is questionable whether this kind of activity could happen at more than a pilot scale, in which case it is open to the criticism that only a limited cosmetic improvement can be achieved.

An interesting test of the voluntary principle is provided by a solution recently adopted for a drainage problem at Halvergate Marshes, a very low-lying area in Norfolk Broadland. The Local Internal Drainage Board proposed to drain 2,388ha of marshland but as a result of negotiation between the Broads Authority, the Internal Drainage Board, the National Farmers' Union, the Country Landowners Association and the Countryside Commission the drainage scheme is proceeding on most of the area while 450ha of the area is left undisturbed and notification agreements apply to the rest of the land. The compromise involves compensation payments to farmers funded to the extent of 25 per cent by the Broads Authority and 75 per cent by the Countryside Commission. The agreement runs with the land for twenty years; and notification agreements will cover agricultural changes. However, the financial commitment by the Countryside Commission is by special Treasury permission 50 per cent over the normal contribution. While this is a valuable pilot scheme, it demonstrates the need for funds for conservation.

The Wildlife and Countryside Act, 1981

The battle in Britain between conservationists, encouraged by such sympathetic government agencies as the Countryside Commission and the Nature Conservancy, and the farming lobby, backed by government in another guise as the Ministry of Agriculture, Fisheries and Food, came to a head in the lengthy debates leading to the passing of the Wildlife and Countryside Act, 1981. Although the numerical strength of the environment lobby is estimated by Goyder and Lowe (1981) at three million people — greater in membership than any political party or trade union — and there is evidence of more sophisticated and professional approaches by pressure groups, the debate has resulted in only slight gains (Shaw, 1981).

Part I of the Act, in dealing with the protection of wild life, merely aligns British legislation with that of Western Europe in general. Part II, dealing with conservation, introduces new and crucial provisions which have the effect of pricing visual amenity. The Nature Conservancy Council and national park authorities have a duty to pay compensation whenever a farmer is refused grant aid by the Ministry of Agriculture on the grounds of conservation in a national park or site of special scientific interest if objections to agricultural development are raised by the statutory conservation bodies. In such cases, the park authority or Nature Conservancy will have to enter into a management agreement with the landowner, to whom compensation is to be payable. While the provision has a superficial attractiveness, there will be no extra funds available to provide compensation; its incidence can scarcely be forecast with confidence; and the overall effect of its financial consequences may be to encourage few objections to the kinds of agricultural change about which the conservationists have become deeply concerned. Significantly, the Act provides no increase in powers for the public acquisition of land, regulating land use, or influencing management techniques. While it firmly recognises that agriculture has a duty towards conservation as well as production, the financial obligations of the public amenity bodies in securing this lie onerously upon them. In these circumstances, the efficacy of the legislation very much remains to be tested.

While the compensation issue has attracted great interest, other provisions of the Act are also noteworthy. The Countryside Commission itself becomes a grant-in-aid body in April 1982 and will no longer be a part of the conventional civil service. District council representation is secured on national park authorities, as a recognition of the tension between local and national needs which the authorities have to resolve.

Given, then, that planning control of the rural areas of Britain is not comprehensive, two questions arise. Firstly, what is the nature of the present constraints? Secondly, what stands in the way of their removal? Seen in a rather simplistic light, the issue may be seen as one of modernised production versus conservation. Activities currently interpreted as 'wealth-producing' — agriculture, forestry or mineral working — clearly belong to the 'production' class, whereas wildlife and landscape conservation are seen by the production lobby as deleterious to their interests. However, the issue is not clearcut, for the production interest does not represent exclusively private enterprise nor does conservation necessarily represent solely public interests. Indeed, competition in the rural scene is broadly analogous to competition in the urban scene in that conservation betokens production or profit foregone.

The questions arise as to how far conservation can be part of the statutory rural planning system and if it is to be so to what extent production or profit foregone in the public interest should be compensated for. A further point is whether conventional allocative land-use-planning techniques are appropriate or whether management planning tools are preferable, relying to a considerable extent on the voluntary principle. Above all, it is perhaps questionable whether the continued interpretation of British rural landscapes according to the National Park — Area of Outstanding Natural Beauty — other rural land hierarchy is still serviceable, given the different statutory provisions for these areas, or whether a planning system more sensitive to highly local issues, rather than the 1949 Act's blanket designations, is needed.

Perhaps the most crucial question in British rural planning in the 1980s is the extent to which the nascent approach to the integrated management of access, services, housing, employment and farming may become a truly coherent planning system. It has influential proponents, particularly in the Countryside Commission, which in the generation of ideas and in local experimentation played a most significant development role in the 1970s. Such an approach would at last begin to resolve the difficulty of the multiplicity of interests, agencies and powers which make development and control of the rural scene so relatively complicated. In brief, forms of plan are needed which will be correct in scale, form and content to reflect the convergence of rural decision-making in the use of a great variety of resources for a great variety of inter-related ends. Only in this way can the negative forces of the statutory planning system and the positive forces of natural resource planning (Blacksell and Gilg, 1981) be harnessed in a constructive relationship.

References

Anderson, M. and Best, R.H. (1981), 'In Search of the Missing Land', *Planning*, 440, 16 October 1981, 8

Barlow Report (1940), 'Report of the Royal Commission on the Distribution of the Industrial Population', Cmnd 6153 (HMSO)

Blacksell, M. and Gilg, A. (1981), *The Countryside: Planning and Change* (Allen and Unwin)

Cherry, G.E. (1976), *Rural Planning Problems* (Leonard Hill)

Clark, D.M. (1981), 'Rural Designated Areas', *Planner News*, May 1981, 8

Clarke, P., Ayton, J.B., and Gilder, I. (1980), 'The Key Settlement Approach', *The Planner*, 66, 4, 98–9, July 1980

Countryside Commission (1970a), *The Planning of the Coastline* (HMSO)

Countryside Commission (1970b), *The Coastal Heritage* (HMSO)

Countryside Commission (1981), *Countryside Management in the Urban Fringe* (CCP, 136)

Countryside Review Committee (1979), *Conservation and the Countryside Heritage*, Topic Paper No 4 (HMSO)

Cullingworth, J.B. (1976), *Town and Country Planning in Britain* (Allen and Unwin)

Davidson, J. and Wibberley, G. (1977), *Planning and the Rural Economy* (Pergamon)

Derounian, J. (1980), 'The Impact of Structure Plans on Rural Communities', *The Planner*, 66, 4, 87, July 1980

Dower Report (1945), *National Parks in England and Wales*, Cmnd 6628 (HMSO)

Essex County Council (1981), *Draft Conservation Plan*

Fairbrother, N. (1970), *New Lives, New Landscapes* (Architectural Press)

Goyder, J. and Lowe, P. (1981), 'Putting on the Pressure', *Planning*, 416, 5

Hall, P. *et al.* (1973), *The Containment of Urban England* (Allen and Unwin)

Heap, D. (1978), *An Outline of Planning Law* (Sweet and Maxwell, London)

Hobhouse Report (1947), *Report of the National Parks Committee*, Cmnd 7121 (HMSO)

Hookway, R.J.S. (1977), 'Countryside Management: the Development of Techniques', *Proceedings of the Town and Country Planning Summer School*, 1977, 7–10

Hookway, R.J.S. (1978), 'National Park Plans', *The Planner*, 64, 1, 20–22, January 1978

Jowell, J. and Noble, D. (1982), 'The Anti Social Plan Modifications', *Planning*, 438, 2 October 1981, 9

Labour Party (1981), *Out of Town, Out of Mind: A Programme for Rural Revival*

MacEwen, A. and MacEwen, M. (1981), *National Parks: Conservation or Cosmetics?* (Allen and Unwin)

McLaughlin, B. (1981), 'Rural Deprivation', *The Planner*, 62, 2, 31–3, March 1981

Martin and Vorhees Associates (1981), *Review of Rural Settlement Policies, 1945–1980* (Department of the Environment)

Moseley, M. (1980), 'Is Rural Deprivation Really Rural?', *The Planner*, 66, 4, 97, July 1980

National Federation of Housing Associations (1981), *Rural Housing: Hidden Problems and Possible Solutions*

Parry, M. (1981), 'The Plight of British Moorland', *New Scientist*, 90, 550–1

Planning Advisory Group (1965), *The Future of Development Plans* (HMSO)

Sandford Committee (1974), *Report of the National Parks Review Committee* (HMSO)

Scott Report (1942), *Report of the Committee on Land Utilization in Rural Areas*, Cmnd 6378 (HMSO)

Shaw, M. (1980), 'Rural Planning in 1980: An Overview', *The Planner*, 66, 4, 88-90, July 1980

Shaw, M. (1981), 'The Wildlife and Countryside Act, Some Lessons for Planners', *Planner News*, December 1981, 9

Shoard, M. (1980), *The Theft of the Countryside* (Temple Smith)

Skeffington Report (1969), 'People and Planning' (HMSO)

Taylor, M. and Bozeat, N. (1982), 'Just How Constraining is the Planning System?', *Planning*, 454, 5 February 1982, 7

TRRU (1981), *The Economy of Rural Communities in the National Parks of England and Wales* (Tourism and Recreation Research Unit, University of Edinburgh)

Uthwatt Report (1942), *Report of the Expert Committee on Compensation and Betterment*, Cmnd 6386 (HMSO)

Wager, J. (1981), *Conservation of Historic Landscapes in the Peak National Parks* (Peak District National Park Authority)

Westmacott, R. and Worthington, T. (1974), *New Agricultural Landscapes*, Countryside Commission, CCP 79

NOTES ON CONTRIBUTORS

Dr D.J. Banister, Bartlett School of Architecture and Planning, University College London, England.

Dr I.R. Bowler, Department of Geography, University of Leicester, England.

Dr M. Bunce, Department of Geography, Scarborough College, University of Toronto, Canada.

Dr A.G. Champion, Department of Geography, University of Newcastle upon Tyne, England.

Dr P.J. Cloke, Department of Geography, St David's University College, Lampeter, Wales.

Dr A.W. Gilg, Department of Geography, University of Exeter, England.

Dr G.J. Lewis, Department of Geography, University of Leicester, England.

Dr M. Pacione, Department of Geography, University of Strathclyde, Glasgow, Scotland.

Professor D.L.J. Robins, Department of Town and Country Planning, Trent Polytechnic, Nottingham, England.

Dr A.W. Rogers, Department of Environmental Studies and Countryside Planning, University of London, England.

Dr M.F. Tanner, Department of Geography, University of Birmingham, England.

INDEX